Lighthouses

Highlights of the Coast

pil

Publications International, Ltd.

Contributing writers: John A. Murray and James Hyland

Images taken from the Library of Congress, the Lighthouse Preservation Society, Shutterstock.com and Wikimedia

Louis Weber, CEO
Publications International, Ltd.
8140 Lehigh Avenue
Morton Grove, IL 60053

Permission is never granted for commercial purposes.

ISBN: 978-1-64030-154-2

Manufactured in China.

8 7 6 5 4 3 2 1

Contents

Introduction

IMAGINE THAT YOU'VE been at sea for two or three weeks. In that time you've made a slow but steady passage across a major ocean. You're on a final heading to a major port. You're making good time, but there is a strong winter storm driving in from the west, and it is blowing at three times the speed of your ship. The winds are building and so are the waves. At first you saw modest whitecaps. Now towering mountains cover your ship with spray and foam, and these monsters are breaking freely across the bow. You are standing on the bridge, envisioning a quiet home where your spouse and child patiently await your return. Home, though, seems so very far away.

All that can be heard is the groaning of the ship and the rain thundering against the windows. The rain pours in sheets across the pitching deck, and every few minutes the foredeck completely disappears under another wave. The early December nightfall is coming fast, and in the last few minutes it has become clear to all on board that the vessel may not beat the storm. There is no turning back at this point. Nor can the ship break north or south to avoid it. The storm is 400 miles wide. The ship is committed to making one port, to finding that single safe harbor on an otherwise unbroken coast.

Suddenly, the rain subsides for just a second, and an eager young mate with sharp eyes sees it—a faint light just to the north. Coastal charts are hurriedly consulted. Someone looks again, this time with powerful binoculars. Yes, there it is— the welcoming light. Half an hour later, your ship passes into the familiar safety of the vast harbor.

Such is the power of a lighthouse to guide a vessel and safeguard a crew after they have made a long journey across the lonely quarters of the open sea. Since the beginning of time, lighthouses have helped ships navigate the often dangerous coasts of the world. In the formative years of this nation, lighthouses were particularly important because the growth of the country was directly tied to commerce with other countries. Foreign trade with France and other friendly nations helped to sustain and save the fragile, young nation.

The first lighthouses in America naturally arose where coastal populations were concentrated and local businesses were growing. Lighthouse building began on the Atlantic Coast, particularly in New England, and then moved south across the Carolinas to Florida and the Gulf Coast, spreading through Alabama, Mississippi, Louisiana, and Texas. The next stage involved the Great Lakes, each the size of some small seas and each requiring the same forms of coastal lighting. In the mid-19th century, lighthouses began to appear on the Pacific Coast. At the same time, lights began to employ French-designed Fresnel lenses, powerful optical devices that focused light through a series of lenses. As a result, the beams could be seen much farther out on the water.

Eventually, lighthouses were built by the federal government on the new territorial possessions of Hawaii and Alaska. Later still, in the 20th century, lighthouses gradually became automated, with remote sensors and automatic timers replacing keepers. Although today only one light station in the United States is permanently manned—Boston Light—lighthouses continue to be vital navigational aids along the coasts of U.S. lakes and seas.

Lighthouses are a permanent part of American history and culture. They are powerful icons, symbols, and metaphors. They are awash in fables and myths, legends and stories. They evoke nostalgia and romance, faith and trust. In a world in which duty has taken on diminished meaning, lighthouses bring to mind a group of people who committed their lives to the absolute responsibility of maintaining the shore light through the darkest hours of the night. In a world of storms, both literal and figurative, lighthouses represent places of safety—sanctuaries from rough seas. In a world that is increasingly crowded and noisy, lighthouses symbolize quiet locations far removed from the turmoil of the cities, and always near the eternal beauty of the sea. In a world of constant change, lighthouses represent places that are permanent.

The journey of life has often been compared to an extended sea voyage. On every voyage there must be a beacon, a guiding light that steadies the final course to an objective. A lighthouse evokes all of those deeper associations and more.

This book celebrates 50 of the most scenic and historic lighthouses. In its pages are saltwater lighthouses and freshwater lights, offshore lighthouses and harbor lights, gigantic 18-story behemoths and diminutive towers a child could climb. Each of these lighthouses was once, or perhaps still is, a faithful star in the night, a beacon that has guided in some lonely mariner from the dark, empty quarters of the sea.

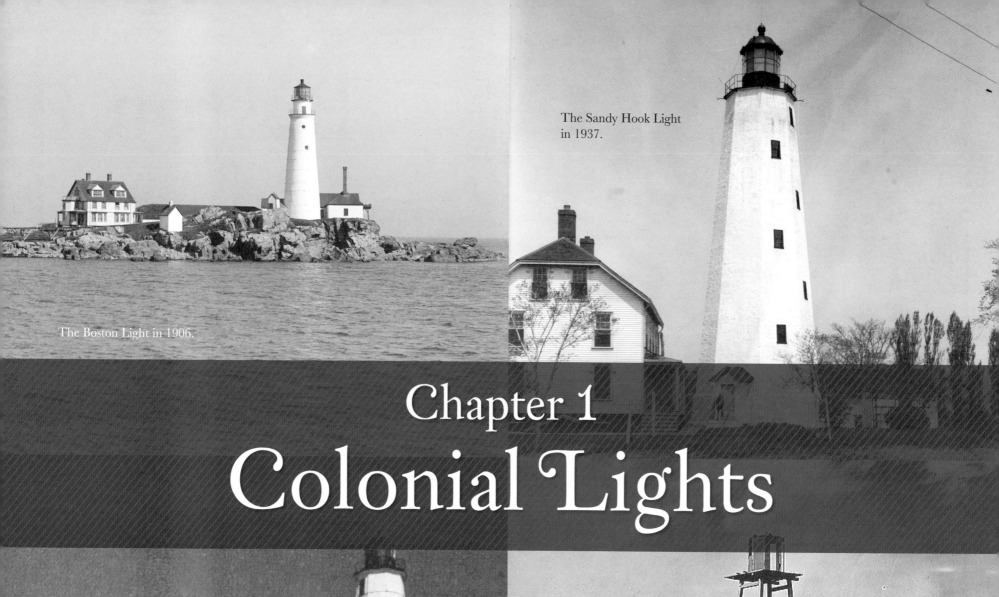

The Boston Light in 1906.

The Sandy Hook Light in 1937.

Chapter 1
Colonial Lights

An undated photo of the New London Harbor Light taken by the U.S. Coast Guard.

The Morris Island Light in ruins with an observation tower built on it during the Civil War.

Boston
Massachusetts (1716)

In the golden light of a summer day, the Boston Light stands watch over one of the busiest harbors in the United States. A short distance to the west, docks await ships that carry flags from virtually every coastal country in the world.

PASSENGERS ON NIGHT flights landing at Logan International Airport often look down into crowded Boston Harbor and see a tiny lighthouse on one of the many islands scattered in Massachusetts Bay. No more than ten miles east of Boston, this light has guided ships toward the port of Boston for nearly three centuries. Just beyond it are some of the busiest docks, piers, and wharves on the Eastern seaboard, if not in the Western Hemisphere. Tankers, freighters, cruise ships, naval vessels, fishing trawlers, recreational sailboats—all have made their way past this historic light en route to one of the most important ports in the world.

Boston, Massachusetts (1716)

Much about the world has changed since the first Boston Light was raised in 1716 (Sir Isaac Newton was still alive in that year!), but through it all—wars, revolutions, social upheavals, changing presidents and Congresses, and scientific breakthroughs—this light has faithfully guided mariners into port.

The Pilgrims had been in eastern Massachusetts for 96 years when they built a lighthouse on Little Brewster Island. "They" in this case refers to the resident merchants of Boston (traders in rum, tea, sugar, cotton, and so on). They successfully lobbied the powers that be in the colony for what historical documents refer to as a "Light House and Lanthorn on some Head Land at the Entrance of the Harbor of Boston for the Direction of Ships and Vessels in the Night Time bound into the said Harbor."

The need for a lighthouse in the early 18th century (before the Industrial Revolution produced steam-powered engines) was even more pronounced than it is in modern times. In the old days sailing vessels made their passages solely at the mercy of the winds and currents, and therefore they could not always leave or make port before nightfall. The lighthouse was seen by merchants as a means of getting loaded ships safely into and out of port more quickly, and their practical arguments proved persuasive. In June 1715 the local authorities appropriated enough money to build a lighthouse on Little Brewster Island at the entrance to Boston Harbor.

Records show the first light appeared at the lighthouse on September 14, 1716, in time for the winter Nor'easters. Hand-drawn sketches that have been passed down from that early Colonial period show a conical, six-story tower with a two-story clapboard lighthouse keeper's residence nearby.

The lighthouse was of such importance that all those who used the waters of Massachusetts Bay were required to pay for its upkeep. History records that the tax rate for incoming freight was one penny per ton to pay for the lighthouse, while outbound freight was taxed at two shillings per ton. All fishing vessels and

This image offers an exciting close-up view of the powerful twin bulbs and second-order Fresnel lens at the heart of the Boston Light.

wood sloops paid a tax of five shillings each year for the lighthouse. The court hired George Worthylake as lightkeeper and paid him a salary of fifty pounds per year. For this he kept the lights burning from dusk to dawn each night. Unfortunately, Worthylake, his wife and daughter, and two other men drowned while returning to the lighthouse two years later. Young Benjamin Franklin wrote one of his first works of literature, an elegiac poem, on this tragedy.

As can be seen here, the Boston Light occupies a strategic point amid the cluster of windswept islands directly east of Boston Harbor and the city.

In 1775 the lighthouse on Little Brewster Island was temporarily shut down to protect Boston from British warships. Disaster struck the following year when the British blew up the lighthouse during their general retreat from the area. It was not until 1783 that the lighthouse was fully restored. The new Boston Light was, like its predecessor, a conical lighthouse, but was given more substantial walls (seven feet thick at the foundation). Its height was raised in 1853 by 15 feet, and it was also given a second-order Fresnel lens in 1859.

Today the Boston Light remains in service—the only lighthouse in the country with resident keepers. In this way the Coast Guard pays tribute to the revered tradition of lighthouse-keeping on the U.S. coasts. Although the island is about ten miles from Boston, it is part of the Boston Harbor Islands National Park and can be viewed on local boat tours and exploration cruises.

The historic Boston Light owns the distinction of being America's first light station. Sandy Hook in New Jersey is the oldest light tower.

Tybee Island

Georgia (1736)

The Tybee Island Lighthouse stands watch over a once remote barrier island, which has become one of the most popular summer vacation spots along the entire Georgia seacoast.

VISITORS TO SAVANNAH, Georgia, one of the loveliest and most historic towns on the southeastern coast, often drive out on the concrete causeway to spend a day at the beach on Tybee Island. Once on the island, these visitors encounter bird-filled salt marshes, miles of sand and sun, and a main boulevard lined with enormous oaks and picturesque sabal palms. On a quiet side street on the north end of the island, near old Fort Screven, sits the Tybee Island Light, a vintage brick structure that towers massively over the surrounding homes and beaches.

The Tybee lighthouse site dates back to the Colonial period. Shortly after General James Oglethorpe established the colony of Georgia in 1732, the local authorities ordered that a lighthouse be constructed on the barrier island. The light would help sailing vessels on their final approach to Savannah. Completed in 1736, the first lighthouse was destroyed by fire only five years later. The second structure was built too close to the shoreline and was threatened by beach erosion. It was then destroyed in 1861 by retreating southern troops shortly after the Civil War began.

Finally in 1867, two years after the war was over, the current structure—an octagonal brick lighthouse—was completed. The bottom 60 feet of the damaged lighthouse were restored, and another 94 feet were added, making the new building some 154 feet in height. At that same time the tower was outfitted with a nine-foot-tall state-of-the-art Fresnel lens that gave the light a visible nighttime range of more than 18 miles. All who saw it knew the welcome entrance to the Savannah River was just to the north of the light.

There is something inspiring, and comforting, about the steady amber light at the top of a lighthouse—for mariners at sea as well as for those who look up from terra firma.

Tybee Island Light is a short and pleasant stroll from nearby beaches. When the Georgia summer sun is merciless, the cool tower on Tybee, which is also the site of a small lighthouse museum, is just the place to go.

The Tybee Island Light has been painted several times since its construction. The tower was restored in 1999 by the Tybee Island Historical Society, who eventually took possession of the tower under the National Lighthouse Preservation Act in 2002.

Brant Point

Massachusetts (1746)

Sunset softly colors the landscape of Nantucket Island around the Brant Point Light. The lighthouse owns the distinction of being the lowest, in terms of elevation, in all of New England.

THERE HAS BEEN a lighthouse on Brant Point, Massachusetts, at the entrance of Nantucket Harbor, since 1746, making this one of the oldest lighthouse sites in the United States. The scene on the island of Nantucket today is very different from that of 250 years ago, when there were only windswept hills of sea grass and scrub oak, with sturdy Atlantic whalers and fishing dories resting in the harbor. Today a myriad of summer homes sit on the gentle hills above the small port, overlooking literally hundreds of private sailing vessels anchored in the waters below. Although not as crowded as nearby Martha's Vineyard, Nantucket is among the most popular vacation sites on the northeastern coast. This easily accessible lighthouse is a favorite tourist destination on the island.

The current lighthouse at Brant Point owns the distinction of being the shortest in New England—its light is only 26 feet above sea level. The building is reached by walking over an elevated wooden walkway across a narrow sand spit. The walkway leads to a pile of rocks on which the tiny lighthouse sits. This tower is the tenth to be built in this area since the first was raised by Colonial mariners in 1746.

In the early days, of course, the major reason for a beacon was to safely guide the fishermen into port. Nantucket was one of the coastal centers for the busy Atlantic fishing and whaling trade in the 18th and 19th centuries. After numerous towers were lost to hurricanes and other violent northeastern storms over the years, a more permanent lighthouse, standing to this day, was finally raised in 1901. Its blinking red light is visible for about ten miles out to sea.

Every summer thousands of tourists travel to Nantucket Island to pay a visit to the lighthouse at Brant Point, walking along this red-painted walkway over the dunes.

Beavertail

Rhode Island (1749)

IN 1749, a bit more than a century after Roger Williams was exiled from Massachusetts and moved west to form the wilderness outpost of Providence, the Colonial authorities of Rhode Island decided to build a lighthouse on Conanicut Island, near the entrance of Narragansett Bay. The area had become well known for its busy maritime traffic, and a reliable lighthouse was needed on the south end of the island to alert sailors that they were approaching the narrow channels leading to Providence. The Beavertail Light, as it was called, would become New England's third, and America's fourth, light station.

Four short years after the original light was raised, it was destroyed by fire and rebuilt. It was burned again, this time by the British during the American Revolution in 1779. The lighthouse remained dark for the rest of the war, but was repaired after hostilities ceased. In 1793 the new federal government assumed control of the Beavertail Light. In 1851 the U.S. Lighthouse Board authorized the construction of a modern new building on the site. Completed in 1856, the square granite tower is still in existence and supports a flashing, white electric light visible about 20 miles out to sea.

Today, the scenic Beavertail Light occupies its own state park, and the assistant-keeper's house is maintained by the state of Rhode Island as a lighthouse museum.

Green algae coats seaside rock formations on this magnificent maritime landscape, which features the historic light at Beavertail.

The Beavertail Light, the country's fourth lighthouse, was built of rubble stone in a square design. It is sturdy and practical, as were its makers.

New London

Connecticut (1760)

There is something reassuring about the last blaze of a lingering sunset over calm waters, accompanied by the soft, amber light of a coastal lighthouse.

THOMAS JEFFERSON WAS just a 17-year-old teenager entering college, and 28-year-old George Washington, recently married, was settling into his role of gentleman farmer, when the first lighthouse was raised at New London, Connecticut, in 1760. Financed by a public lottery, the original lighthouse was erected on the heavily forested southwestern bank of the Thames River, where the river empties into the northern waters of Long Island Sound. Here its light could be plainly seen by passing vessels as a warning to captains that if they sailed further, they would miss the Thames and the port of New London.

Thirty years later, after the Revolution, control of the lighthouse was taken over by the federal government. Eventually, in 1801, the Colonial lighthouse was replaced by a massive octagonal tower that rose 80 feet above the surrounding waters. Attached to it was a lovely three-story home made of the same brick as the light tower. In 1855 a state-of-the-art Fresnel lens was installed in the tower. The 1801 tower remains today, although it is now privately owned and closed to the visiting public.

Trident nuclear submarines, based at New London Naval Submarine Base in nearby Groton, now churn through the same Thames River waters that were once familiar to Atlantic whalers and cod fishermen in the Colonial era. Much about the world has changed since the days of Jefferson and Washington, but one thing has remained constant—the need for coastal lighthouses like the old tower at New London to protect the lives of recreational, commercial, and even naval mariners.

Rising just at water's edge, the lighthouse at New London is one of the taller lights in New England. From base to light it rises some eight stories above ground level.

Sandy Hook

New Jersey (1764)

If these stones could talk, they would tell an incredible story of bright dawns and fiery dusks, peaceful days, and storms that went on for days.

LOOK AT A MAP of the greater New York City area and, about ten miles south of Manhattan, you will see Sandy Hook, a narrow spit of New Jersey that is part of the Gateway National Recreation Area. On that sandy, windswept peninsula is the Sandy Hook Light National Historic Landmark. The Sandy Hook Light is the oldest standing lighthouse in the United States. It also owns the distinction of being the country's first octagonal lighthouse. The octagonal design, which became very popular in the 19th century, was created to ensure a more durable structure that would be better able to weather severe storms. Near the tower is a brick, three-story keeper's house, complete with veranda and two chimneys.

The Sandy Hook Light in New Jersey, constructed in 1764, is still operational today, making it the nation's oldest continuously working lighthouse.

In the early 19th century, the lighthouse employed a large whale-oil lamp that had four-dozen wicks. It was eventually modernized in 1856, when a Fresnel lens and state-of-the-art lamp replaced the whale-oil light. Interestingly, the classic Fresnel lens remains in place today and radiates a continuous beam of light that is visible well into New York City. The Sandy Hook Light has its share of ghost stories. One legend has it that builders found a skeleton behind a makeshift brick wall!

In the early 21st century, as in the mid-18th century, all a ship's captain has to do on final approach to New York is steer to the north of the Sandy Hook Light and he or she will reach the city's safe waters. For well over two centuries, a light has burned every night on the top of this classic 85-foot stone tower, a remarkable testament to the well-built soundness of this Colonial lighthouse, as well as to the enduring importance of these navigational aids.

Plymouth
Massachusetts (1769)

The striking pyramidal shape of the light at Plymouth, a very sturdy form of lighthouse design, can be observed in this photograph.

THE HISTORIC PLYMOUTH Light occupies a thumb-like projection of land near the entrance of Plymouth Harbor, about 30 miles south of Boston. Although this flat point of land, called the Gurnet by local residents, is today crowded with fine seaside homes, it was once a lonely, desolate peninsula far from human habitation.

The first light was built in 1769, nearly 150 years after the Pilgrims, sailing from England aboard the *Mayflower*, initially settled around Plymouth Harbor. The original lighthouse consisted of two beacons, one at either end of a small building. It had the unusual distinction of having been hit by a British cannonball during the American Revolution, although the light was miraculously undisturbed.

After the Revolution was over, the lighthouse was ceded from Massachusetts to the new federal government. Its tradition of having two lights at the point persisted into the 19th century in order to avoid confusion with other nearby lights. At least two replacement structures also contained a pair of beacons. The last two-beacon edifice was raised in 1843, although one of the twin towers was later taken down in 1924. Its 39-foot mate still remains, flashing every 30 seconds.

Clearly visible from tour boats in Plymouth Bay, which remains a popular vacation site because of its Colonial importance, the Plymouth Light is situated on high sandy bluffs. It is a wonderful place to take the family and reflect upon the historical importance of Plymouth, where the first pioneering vessels from England made landfall in New England, where the concept of freedom took a step forward, and where a pair of faithful lights welcomed mariners in from the sea.

The first Plymouth Light was constructed seven years before the Declaration of Independence was signed in Philadelphia.

Portsmouth Harbor

New Hampshire (1771)

On this bright, sunny day, the cobalt blue waters off Portsmouth are filled with recreational sailboats, evoking a painting by that master of maritime artists, Winslow Homer.

The green, flashing light of the Portsmouth Harbor Light still helps mariners find their way near the entrance of the Piscataqua River.

LEGEND HAS IT that none other than President George Washington came to New Hampshire to visit the Portsmouth Harbor Light in 1789. He found its general maintenance to be lacking, and arranged for the lighthouse keeper to find another line of work. By that time the lighthouse, located in New Castle at the entrance of the Piscataqua River, had been in operation for 18 years. During the Revolutionary War, it helped guide the local New England privateers who raided British shipping and harassed the king's Navy vessels. Local history also has it that, during the war, folk hero Paul Revere and local militiamen captured the fort in Portsmouth Harbor with the help of the lighthouse keeper.

In its early days, the light was quite primitive—little more than a lantern hung from a flag mast at the harbor fort. By 1800 the old lighthouse was in a state of ruin, and a new one was planned. Four years later the original tower was replaced with a new one that rose from a foundation of granite blocks. A Navy report in 1838 stated that the Portsmouth Harbor Lighthouse was in "fine order."

Only 28 years later, in 1877, the U.S. Lighthouse Service replaced this 80-foot tower with a 48-foot cylindrical iron tower. That tower, which emits a fixed green light, remains today and is said to be visible from about 12 miles out. Although the Portsmouth Harbor Light is currently part of a U.S. Coast Guard base, it is open regularly for public visitation. The lighthouse is still on the grounds of the old Colonial fort that it once served. Those who pay a visit to Portsmouth Harbor Light can join the ranks of its other visitors through the years, which include New Hampshire native son Daniel Webster and the widely wandering Henry David Thoreau.

Cape Ann

Massachusetts (1771)

These two lights on Thacher Island at Cape Ann have stood watch over the seas since 1771. They were among the first lights on the northeastern coast.

THE CAPE ANN LIGHT, located on Thacher Island about 40 miles north of Boston, is a lighthouse with some interesting history. First of all, there is the matter of the island itself—its name comes from Anthony Thacher, who was given the island in 1635. He and his wife had been the sole survivors of a shipwreck there in which 21 people—including the Thachers' four children—lost their lives. Cape Ann's first lighthouse keeper, Captain Kirkwood, was removed from his position during the Revolutionary War because he was a Tory, a supporter of the British.

In 1919 the foghorn attached to the light-house at Cape Ann probably saved the life of President Woodrow Wilson. Wilson was on his way back from Europe following the Versailles Peace Conference, which ended the First World War. His ship, the S.S. *America*, became lost in the fog and was inadvertently heading right for Thacher Island. Just before an unthinkable catastrophe occurred, the Thacher Island foghorn alerted the captain to the ship's perilous position and he was able to change course.

The Cape Ann Light is among the oldest lighthouses on the coast, having first become operational in 1771. Since its earliest days it has been a twin lighthouse—that is, it is made up of two lighthouses separated by about 300 feet on the small rocky island. The two 124-foot granite towers on Thacher Island date to 1861. The south tower remains active, producing a flashing red beacon. The north light is operated as a solar-powered amber light, but only as a memorial to mariners past and present. Thacher Island, which is accessible only by boat, can be seen from the state highway near Rockport on Cape Ann.

There is something noble and inspiring about a line of breakers crashing into a rocky headland. Across the waters, rising like a vision from the windswept waters, are the twin lights of Cape Ann.

Morris Island

South Carolina (1767)

There is a movement afoot in the city of Charleston to restore the light at Morris Island, primarily because of its importance to the history of the city and the region.

ARCHIVAL U.S. COAST GUARD photographs, circa 1885, show South Carolina's Morris Island Light surrounded by dense subtropical trees. Adjacent to it is a stately keeper's home built in the grand Victorian style. This three-story wood-framed residence is painted white with dark shutters, and four red-brick chimneys rise from its steep, shingled roof. Long wooden boat docks project out into the waters of Charleston Bay.

Today all of that has changed completely, as the derelict lighthouse stands alone, with no island around it, about 300 feet from the nearest

The interior of the current Morris Island Light has reverberated with the sounds of hundreds of hurricanes. Yet the tower still stands after more than 120 years.

shoreline. Despite having weathered more than 100 hurricanes and several major earthquakes, the Morris Island Light is now threatened by simple beach erosion. The nearby Charleston Light on Sullivan's Island has lit the way for mariners in the area since 1962.

Those who swim or paddle out to the weathered Morris Island Lighthouse will find on the foundation a cornerstone with a copper plate that reads: "The first stone of this beacon was laid on the 30th of May 1767 in the seventh year of his majesty's reign, George III." The current tower was actually built in 1876, after a century of preceding lighthouses on the site had come and gone. Standing over 160 feet tall, the existing tower rests on a thick concrete foundation that in turn is set atop deep piles in the mud and sand of the harbor floor. This partially accounts for why the structure still remains, despite the severe earthquake that ruined much of Charleston in 1886. It is now what navigators call a "day mark"—a structure that can be seen fairly well on a clear day and can alert passing vessels to the nearby beach.

Recently, a local organization called Save the Light purchased the Morris Island Light and then sold it in 2000 to the South Carolina Department of Natural Resources for $1. The state has since leased the light back to the preservation group. Work to stabilize the lighthouse began in 2002.

The Morris Island Light was automated in 1938 and was deactivated in 1962. The original lens was a first-order Fresnel lens which was cracked in the 1886 Charleston earthquake.

Throughout the years, the buildings that have surrounded the Morris Island Light have been completely destroyed by the numerous hurricanes the island has sustained. Hurricane Hugo in 1989 razed all of the remaining structures except for the tower itself.

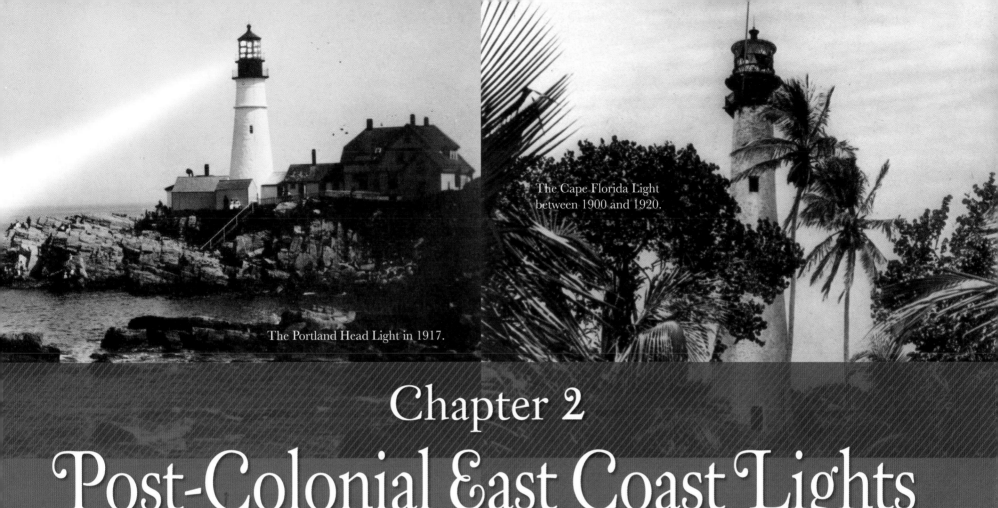

The Cape Florida Light
between 1900 and 1920.

The Portland Head Light in 1917.

Chapter 2
Post-Colonial East Coast Lights

The Minot's Ledge Light
between 1880 and 1899.

The old (right) and
new (left) Cape Henry
Lights in 1905.

Portland Head

Maine (1791)

Although the lighthouse keeper's residence has been modernized, the scene at the Portland Head Light has remained essentially the same since 1790, a tribute to the foresight of Colonial planners.

FOR NEARLY A CENTURY, the bulk of the nation's lighthouses were found along the shores of the New England colonies (later states). As late as 1830, more than two-thirds of America's lighthouses were clustered in the 600-mile stretch of coastline between New York and Maine. The reasons for this heavy concentration of lighthouses in New England are clear: Most of the burgeoning maritime trade with France and other European countries came in and out of ports in this area.

At the time, the southern colonies operated from just a handful of scarce deep-water ports, while in New England deep-water ports with easy ocean access abounded. The southern states were more exporters (cotton and agricultural products) than importers and traders. New England was already exploding in the Industrial Revolution.

The stairway inside the Portland Head Light reveals the craftsmanship of its builders. Note the careful attention that was given to the fine wrought-iron work.

Portland Head, Maine (1791)

The Portland Head Light has marked the entrance to the busy harbor of Portland since 1791. It is one of the most beautiful lighthouses in New England, if not the entire country, and is one of the major tourist attractions along the Maine coast. Both the keeper's house and the 80-foot cylindrical tower sit upon a massive point of weathered rock near the entrance to Portland Harbor. The picturesque scene—wild surf, broken rock, immaculate lighthouse—is a favorite of marine oil painters and landscape photographers.

The Portland Head Light holds the distinction of having been dedicated by the Marquis de Lafayette, the distinguished Revolutionary War general from France. It was the first lighthouse to have been built by the United States Government (as opposed to a local Colonial authority).

The bill setting out its establishment was passed by the new United States Congress on August 7, 1789. Then President George Washington instructed his secretary of the treasury, Alexander Hamilton, to finish completing what had only been modestly begun by the old Massachusetts colony. The total cost of constructing the new lighthouse was $3,000, a regal sum for that time period.

In 1855 the tower received a modern Fresnel lens, as well as a masonry lining and an interior circular stairway made from wrought iron. In 1885 the tower was raised 20 feet, and in 1891 the keeper's house was enlarged and a garage and oil house were added. Other than that, the lighthouse

The lighthouse keeper's residence at Portland Head, with its green gables and steeply pitched red roof, is a thing of architectural beauty.

The shoreline around the Portland Head Light alternates between rugged rock formations and beaches of smooth, sorted cobblestone. Portland Head is one of the most visited lighthouses in America.

looks pretty much the same as it did when General Lafayette came up for the dedication more than two centuries ago. Maine has more than 2,000 miles of coastline crammed into the various bays, inlets, coves, islands, islets, and channels along its eastern shore, and there are more than 60 lighthouses from north to south. Of all those lighthouses, though, none are more lovely, or more easily seen, than Portland Head. This is the light that inspired Henry Wadsworth Longfellow to write his lovely poem "The Lighthouse." The powerful beacon is said to be visible for up to 25 miles. Visitors today will find a wonderful lighthouse museum in the old keeper's residence.

This historic bell sits at the base of the tower at Portland Head, evoking a time before radar and radio, when bells were used as signaling devices.

Cape Henry

Virginia (1792)

HISTORICALLY, the lighthouse at Cape Henry has been one of the most important navigational aids along the mid-Atlantic coast. Cape Henry is located in southeastern Virginia at the southern entrance of Chesapeake Bay, an important saltwater fishing region as well as a natural corridor for commercial shipping. As early as 1721 the governor of the Virginia colony was actively promoting the idea of a lighthouse at that location. Getting all involved parties to agree—Colonial and British authorities as well as private companies—proved an insurmountable obstacle, however. The Ninth Act of the First Congress, established in 1789, stated that the government would build this important lighthouse. It was not until 1792 that a lighthouse was finally constructed on the Cape Henry site.

A full moon rises slowly from the water surrounding the Cape Henry Light. At such times, tides run full and fishermen can find rich bounties in the sea.

The 90-foot tower, built of locally quarried sandstone blocks, has stood on the spot for over 200 years, weathering all manner of hurricanes and ocean-born storms and squalls. The Cape Henry Light was outfitted with a modern Fresnel lens in 1857 and, during the same period, its interior was reinforced with a layer of bricks. After being slightly damaged by Confederate troops early in the Civil War, the lighthouse was returned to duty in 1862 by Union forces.

Nineteen years later, in 1881, a second tower was built near the original Cape Henry Light. That cylindrical tower, made of cast iron, stands 163 feet tall. The new lighthouse was fitted with a first-order Fresnel lens that casts its light far out on the waters of the bay. The original Cape Henry Light still stands, and can be seen near its replacement tower.

One of the quiet joys of a lighthouse is walking the dunes, fields, and beaches nearby. The seas can be a source of solace and renewal.

Montauk Point

New York (1797)

The seas surrounding the Montauk Point Lighthouse were historically one of the richest fishing regions in the United States.

Morning light illuminates the massive octagonal light tower at Montauk Point. In the foreground is the three-story lighthouse keeper's residence, with a decommissioned ship's mast near the porch.

EAST OF FIRE ISLAND, east of the Hamptons, and east of every other geographic point on Long Island is Montauk Point. There, more than 100 miles from New York City, is one of the first points of land mariners have historically observed as they sailed to that great capital of maritime commerce. Ordered by President George Washington and designed and built in 1797 by preeminent architect John McComb, the 78-foot tower at Montauk Point rises from a grassy headland to overlook the gray-blue Atlantic. From there, with saltwater on two sides of the windblown hill, visitors can observe wheeling gulls, passing whales, and the constant stream of oceangoing traffic heading in and out of New York City. Given its location on tall bluffs above the surf line, Montauk Point is one of the most scenic lighthouses in North America.

Montauk Point, New York (1797)

The octagonal tower that McComb constructed was built of sandstone blocks quarried on Long Island. Together with the adjacent two-story lighthouse keeper's residence, the Montauk Point light station cost over $22,000, a sizeable fortune at the time. Given the importance of foreign trade, however, particularly with France, it was a sound investment for the early republic. The station's light, standing on top of a tower that itself rests on an 80-foot bluff, beams out from a height of nearly 170 feet, which makes it visible far out to sea.

Today the Montauk Point Light sees over 100,000 visitors each year, partly a result of the fascinating Montauk Point Lighthouse Museum. The museum displays a large collection of lighthouse lamps and lenses, as well as an interactive diorama designed to acquaint children with the function and importance of lighthouses throughout U.S. history.

The Montauk Point Light was the first public works project completed by the U.S. government, the first lighthouse constructed in New York, and the fourth oldest active lighthouse in the nation.

The current lamp flashes every five seconds and can be seen from nearly 17 nautical miles away.

The Montauk Point Light has been listed on the National Register of Historic Places for many years, but it was only recently added as a National Historic Landmark in 2012 for its importance to New York's international shipping industry.

Cape Cod

Massachusetts (1797)

Though it's a grueling climb to the top of the 66-foot-high lighthouse, the walls have provided unbeatable protection against the many violent storms that have slammed the coast.

"THIS LIGHTHOUSE, known to mariners as the Cape Cod or Highland Light, is one of our 'primary sea-coast lights,' and is usually the first seen by those approaching the entrance of Massachusetts Bay from Europe." So wrote renowned Massachusetts naturalist Henry David Thoreau in 1855, after one of his several visits to Cape Cod. Although Thoreau reported in his essay that the resident lighthouse keeper believed the Cape was in danger of being overrun by the sea, both the narrow sandy peninsula and the sturdy lighthouse are still standing, a century and a half later.

Today, deep within the borders of the 43,000-acre Cape Cod National Seashore, the Cape Cod Light is among the most important historic sites on the peninsula. The lighthouse is the oldest on Cape Cod, having been established in 1798. The tower is 66 feet tall, occupying a sandy bluff that rises more than 130 feet above the beaches. Consequently, its light is visible far out to sea. Over the years the lighthouse has seen the cliffs on which it sits increasingly erode. It was rebuilt in 1857 and then, in 1996, it was saved from erosion by engineers, who moved it back 450 feet.

Thoreau once spent a night with the lighthouse keeper and reveled in the keeper's stories of great storms. Having been replaced by an array of sensors and timers and by a powerful flashing beacon, the faithful keeper and his Fresnel lens are no longer present. Still, the historic tower and the splendid maritime scenery remains as Thoreau knew it. The Cape Cod Light is one of the most beautiful on the northeast coast. It should be visited by all those who love lighthouses and who appreciate the rich history of Cape Cod.

Cape Cod is famous for its quick-developing fog banks, as the warm waters of the Gulf Stream meet the colder air of the far north. At such times, the Cape Cod Light can be a real lifesaver for mariners disoriented by atmospheric conditions.

The Cape Cod Light was the first lighthouse to be built on Cape Cod in order to warn mariners about the dangerous coastlines found between Cape Ann and Nantucket.

Cape Hatteras

North Carolina (1803)

The last golden light of day breaks across the beaches of Cape Hatteras. On this perpetually windy, exposed point, the eternal battle between the land and the sea can be clearly observed.

IN THE NATIONAL ARCHIVES, there is a pen and ink sketch made by Brigadier General George Nicholson on November 9, 1870, of the Cape Hatteras Light in North Carolina as it existed in that year. The sketch shows the Cape Hatteras Light pretty much as we see it today, a towering octagonal structure rising massively above the beach and surrounding sand hills. The simple drawing evokes much of the beauty and power of this fabled tower. The Cape Hatteras Light has now existed, in one form or another, for nearly two centuries and is probably the most famous lighthouse in the United States. It is both a visual icon and a moving symbol of humanity's eternal battle with the fury of the sea.

Several factors have contributed to the importance of the Cape Hatteras Light over time, all of them related to its unique geographic location. First, the Cape Hatteras region—a 300-mile-long array of barrier islands called the Outer Banks—juts far out to sea, putting its fragile lowlands, quite literally, in the middle of the Atlantic Ocean. Second, the eastern continental shelf, a natural protective barrier, is shorter along the Outer Banks than along the Georgia and Florida coasts. Third, a steady stream of tropically born hurricanes annually tracks north on a collision course with the Outer Banks. Fourth, just offshore from Cape Hatteras, the warm waters of the Gulf Stream meet up with the cold waters of the Labrador Current, creating a deadly mix that can quickly produce sudden squalls and violent storms.

It is a long walk up these steep stairs to the top of the Cape Hatteras Light, which is the tallest lighthouse, from ground to roof, in the United States.

The combined effect of all these factors has caused mariners for centuries to call the seas and shoals around Cape Hatteras the "Graveyard of the Atlantic." Records show that over 2,000 ships, both large and small, have sunk in the area since Colonial times, many in the dreaded Diamond Shoals, an enormous underwater sand bank projecting more than 12 miles eastward from the cape. Although Congress authorized the Cape Hatteras Light in 1794, political infighting dragged the construction process on for years. It was not until 1803, during the administration of Thomas Jefferson, that the first 95-foot stone tower was raised. Unfortunately, it was not tall enough to cast a light beyond the treacherous Diamond Shoals. When the Cape Hatteras Light was formally inspected by the U.S. Lighthouse Board in 1851, it was described as "the worst light in the world."

Cape Hatteras, North Carolina (1803)

After the Civil War, the federal government recognized the necessity of raising a new, modern light at Cape Hatteras Point. A 193-foot lighthouse was built on a solid eight-sided foundation with deep iron and timber pilings set into the sand and clay. It was completed in 1870, just as national and international seaborne commerce was beginning to return to the war-torn ports of the Old South. Although it has weathered more than 100 hurricanes in its 130-year history, the Cape Hatteras Light, which once stood more than 400 yards from the breaker line, was, until recently, at the ocean's edge, a result of steady beach erosion. Both the North Carolina state government and the federal government worked to preserve America's most beloved lighthouse, and on July 9, 1999, it was moved back from the eroding beach at a cost of $10 million. Given the importance of this light, their success seems assured.

Part of the fun and adventure of viewing the Cape Hatteras Light resides in the journey to get there. A trip to the light begins on the mainland, near the small sunburned-tourist community of Manteo on Roanoke Island. Visitors drive across the causeway to just below Nags Head, where the road turns south along the windswept landscape of the Outer Banks, the sea on one side of the asphalt road, the grassy sand dunes on the other. A journey of some 50 miles, every inch of which is like something from a Winslow

Oftentimes we lose sight of just how large a light tower can be. The man on the catwalk provides some sense of perspective for the size of the light room at the top of Cape Hatteras.

In this expansive aerial view, the environs of the Cape Hatteras Light can be fully appreciated. Notice how far out to sea the breakers occur near the point of Cape Hatteras, which is indicative of dangerous offshore sand shoals. The lighthouse is visible on the upper right.

Homer painting, ends at Cape Hatteras, with its giant spiral-striped lighthouse looking resolutely over the sea.

On a clear day, sightseers with binoculars can spot the Diamond Shoals Lighthouse, which sits like an offshore drilling platform directly over the perilous Diamond Shoals. Visitors to the Cape Hatteras Light will also find a delightful visitor center and a museum of Outer Banks maritime history in the old lighthouse keeper's residence near the light.

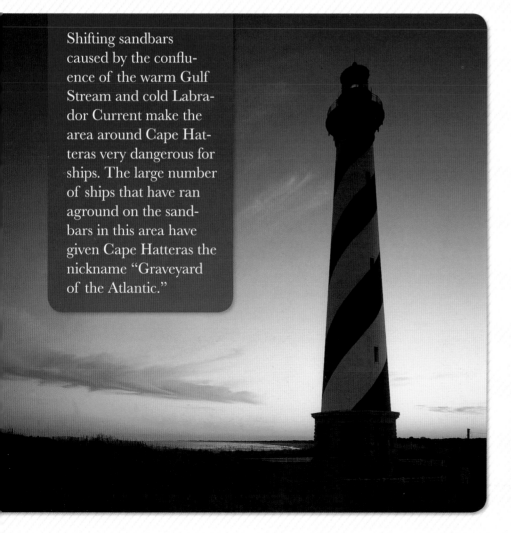

Shifting sandbars caused by the confluence of the warm Gulf Stream and cold Labrador Current make the area around Cape Hatteras very dangerous for ships. The large number of ships that have ran aground on the sandbars in this area have given Cape Hatteras the nickname "Graveyard of the Atlantic."

This photograph shows the massive amount of construction equipment assembled around the Cape Hatteras Light, as efforts were made to protect the tower from the rapidly eroding beachfront.

West Quoddy Head

Maine (1808)

The West Quoddy Head Light rises distinctly from the surrounding wildflower-filled fields with parallel red stripes. From a distance, it resembles a barber pole or a gigantic piece of candy.

THE WEST QUODDY HEAD LIGHT in extreme northeastern Maine is to the northern Atlantic seaboard what the Admiralty Head Light in extreme northwestern Washington is to the northern Pacific seaboard—the most remote station on the wildest outskirts of that coast. Established in 1808, the West Quoddy Head Light is one of the nation's oldest and most revered lighthouses. For nearly two centuries it has lit the often-stormy channel between the United States and Canada. The West Quoddy Light is located on a point of land known as West Quoddy Head, just across the long blue-gray channel from the southern shores of the Canadian province of New Brunswick.

The original West Quoddy Light, built of local stone, was dismantled in 1857 and replaced the following year with a 45-foot brick tower. At the same time, it was outfitted with a Fresnel lens. With its distinctive red and white markings, the tower stands out on the headlands. Its pulsing light can be seen every 15 seconds for about 15 miles on a clear day. The station is also equipped with a resonant foghorn, although this was not present from the beginning. Earlier fog signaling devices included an enormous bell, a steam whistle, and a sea cannon. An important historical feature of the light at West Quoddy Head is the bell, which was America's first fog bell.

The heavily wooded atmosphere around West Quoddy is reminiscent of the Canadian far north, rather than its American New England location. Although West Quoddy Head Light is remote, it is easily visited, as it is part of its own state park.

The park offers nature trails, panoramic views of the channel, and the opportunity—always tempting—to photograph or paint the beautiful lighthouse. For those who come near dusk, beware: Legend has it that the lighthouse is haunted.

Conical and cylindrical lighthouses, such as the West Quoddy Head Light, are among the most common along the eastern coast.

Cape Florida

Florida (1825)

THE FLORIDA KEYS are one of the loveliest landscapes in the United States: the salt-water shallows with their infinite shadings of azure and turquoise, the sun-washed white sand wherever an island rises above the water, the great purple thunderheads that can rise on the horizon at any time of the year....From Miami, the

Experienced mariners know that the green waters surrounding the Cape Florida Light are quite shallow (deeper waters are always a deeper shade of blue). Hence, the need exists for a good light that can be seen from quite a distance out to sea.

keys run south and west for nearly 200 miles, past Elliot Key and Key Largo and Indian Key, until finally, at Hemingway's Key West, they are within only 90 miles of the Cuban coast. This is the sort of place where every third person has a boat of some sort, and most people take recreation on the ocean as often as those who live near the mountains seek solace in the peaks.

From the beginning, the Florida Keys have posed a number of serious problems for mariners. This was true even in the earliest days of exploration during the 16th century, when Captain Rene Laudonierre was exploring the area for France and his vessel was unceremoniously wrecked on the south Florida coast. If the Florida Keys pose navigational problems during calm weather, the situation becomes even more dire during the numerous hurricanes that beset the keys every year. Nowhere was the need for a reliable lighthouse more pressing in this area than at the head of Biscayne Bay, just south of where Miami is located today.

It was in 1825 that the first lighthouse was built at Key Biscayne, shortly after the United States acquired Florida from the Spanish. The lighthouse was seen both as a navigational aid for Biscayne Bay, since trade was burgeoning between the Old South and Latin America, and as part of a system of coastal defenses against the pirates and political turmoil of the Caribbean, traditionally known as a dangerous area. The lighthouse consisted of a conical brick tower, generally painted white, and a black light room.

At the heart of every lighthouse is the light and lens assembly. Working together, they project the beacon far out onto the surrounding waters.

Not long after it was raised, during the Second Seminole War, the Cape Florida Light was attacked by Seminoles in one of the most dramatic incidents in the history of American lighthouses. The assault occurred on July 23, 1835. In mid-afternoon on that day, approximately 40 warriors carrying muskets, with their faces painted with war paint, moved up through the mangroves. The lighthouse keeper, John Thompson, and his assistant moved into the brick tower and locked the heavy wooden door. Armed with muskets, they managed to defend the lighthouse for quite some time, but finally the warriors set fire to the door and window. As the wooden tower steps burned, the keeper and his assistant sought refuge in the lens room. As the flames climbed higher through the wooden tower steps, Thompson's oil-soaked clothing began to burn. Thompson later recalled that he thought the end was near. The Seminoles, perhaps assuming they had killed the men, retreated.

Cape Florida, Florida (1825)

A naval vessel, the USS *Motto*, patrolling nearby saw the flames. Its crew set out to save Thompson, who had been seriously burned and was stranded in the tower room. His assistant had been struck by a musket ball and had not survived the attack. The act of saving Thompson required some ingenuity on the part of the sailors because, with the tower steps burned away, there appeared to be no way to reach him. Finally, someone came up with the idea of shooting a line up to the tower room. Thompson then hooked the line to the railing, and two sailors were able to climb up. Using a makeshift litter, they lowered Thompson to the ground.

The Cape Florida Lighthouse was not reconstructed during the next decade because of the constant threat of attack by the Seminole Indians. Finally, in 1846 work began on the lighthouse tower. By 1855 it was restored to full operation and

The historic residence for the lighthouse keeper at the Cape Florida Light is a sturdy red-brick structure. Even in the Keys, the winter nights can be cool and damp, as evidenced by the two chimneys at either end of the house.

raised to 95 feet. The Cape Florida Light burned brightly until the Fowey Rocks Lighthouse was raised on an open-water reef nearby. For more than a century, the Cape Florida Light stood abandoned and lightless.

In 1996, however, the lighthouse was restored by the Dade County Heritage Trust, which is devoted to maintaining historic buildings in the greater Miami area. The Cape Florida Light, surrounded by graceful royal palm trees imported from the Caribbean, now forms a popular tourist attraction at the Bill Baggs Cape Florida State Recreation Area.

It was through this small window in the Cape Florida Light that the beleaguered lighthouse keeper, in 1835, stared out helplessly at the surrounding band of Seminoles.

This is how the Cape Florida Light looked before it was painted white several years ago. The stone jetty visible to the left of the tower helps to keep the constant erosional forces of the sea at bay.

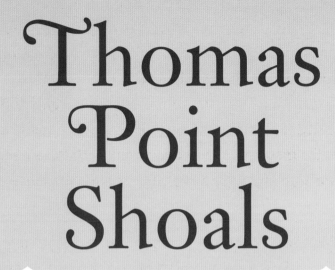

Thomas Point Shoals

Maryland (1825)

The Thomas Point Light was automated in 1986 and was sold to the city of Annapolis, Maryland in 2004. The structure is maintained by various local and national organizations.

GEOGRAPHERS HAVE OFTEN described the 190-mile-long Chesapeake Bay as the remnant of an immense river valley. Once the basin of the Susquehanna River, this lowland area in Maryland gradually flooded as ocean waters rose at the end of the Ice Age, around 15,000 years ago. What this means for recreational and commercial vessels today is that the Chesapeake Bay waters are often very shallow, and the bottom is covered with thick and potentially dangerous shoals of mud, sand, and silt. If a ship veers even slightly off course from one of the main channels, it can find itself stranded far from shore on a flat shoal just below the surface. In even a modest storm, this can be a fatal situation for the captain and crew. Hence, there is a serious need for reliable lighthouses in Chesapeake Bay, both onshore and offshore, to indicate the presence of these hazards.

The light at Thomas Point Shoals, located near Annapolis, Maryland, is one of the more unusual-looking lighthouses in America. The Thomas Point Light is perched on a wrought-iron platform in the middle of Chesapeake Bay. The light's base was created with a type of construction that secures the building to the ocean floor, allowing it to sit surrounded by water. The light indicates the presence of the treacherous shoals upon which Thomas Point Light is located.

The light and the two-story lighthouse keeper's residence are part of the same structure. It is a thing of beauty—hexagonal in shape, cottagelike in appearance, with a porch running completely around the first floor. The roof is painted bright red, and the light room, painted black, rises from the top of the whole affair. In former times, beginning in 1825, the Thomas Point Lighthouse was located onshore. In 1875 a new Thomas Point Light was moved to its present position offshore, where it was found to be more useful. Its flashing beacon still helps tankers, freighters, fishing boats, and small recreational craft stay clear of the Thomas Point shoals.

Thomas Point Light is the only "screwpile" lighthouse still remaining in Chesapeake Bay. Curious passers-by love to gawk at the unusually designed light.

Navesink

New Jersey (1828)

IN 1609 CAPTAIN HENRY HUDSON, an English explorer attempting to discover a northeast passage to the Far East, came upon the mouth of the great river that now bears his name. Before he entered the mouth of the river, sailing as far north as the place where Albany sits today, he took note of the unusual hills along the northern coast of what we call New Jersey.

What Henry Hudson established was that the uplands of the northern New Jersey coast were helpful aids

The late afternoon sunlight shines brightly on the main tower of the Navesink Light. The lighthouse is as much an architectural accomplishment as it is an engineering achievement. Notice, for example, the fine metal detail on the circular roof of the light.

Seen from this frontal perspective, the length of the Navesink Light presents the appearance of a castle, with the two light towers balancing the whole.

in determining one's position relative to the mouth of the Hudson River. This fact did not escape the attention of subsequent mariners in general and, some years later, of the United States Government in particular. So it came as no surprise to anyone that, in 1828, the federal government built a light station on the commanding 200-foot New Jersey hills that Hudson had first surveyed. To help distinguish the Navesink Light from others to the north and south, officials erected two octagonal towers, which were separated by a distance of about 320 feet.

The Navesink Lights were rebuilt in 1862. The south tower was given a square shape, while the north tower was octagonal. Centered between them was a fortress-like building, with huge flanking walls extending to either tower. The whole massive affair was made of brownstone and resembled a gigantic medieval castle, complete with turrets. All of this elaborate design was intended to help mariners distinguish the Navesink Light from the Sandy Hook Light a few miles to the north. Any confusion could prove fatal, as a ship captain might turn his vessel toward the west prematurely, thinking he was approaching New York, and wind up grounded off the New Jersey coast.

Navesink, New Jersey (1828)

In 1898 the north Navesink tower was decommissioned, and the south tower became the first lighthouse in the U.S. to use an electrically powered lighting device. The south tower continued serving mariners until 1953, when it was also decommissioned. Today only a small, memorial-like beacon remains on the south tower, paying quiet homage to the historic role these twin lights played in developing and protecting commerce along the eastern seaboard.

One little known but fascinating fact about the Navesink Light is that it was adopted as the official symbol of the American Army Corps of Engineers, who constructed it in 1862. For well over a century, the Navesink Light has been worn as an emblem on that service's jackets and caps.

Few lighthouses in America have been constructed as sturdily as the Navesink Light. Here we see its massive brownstone blocks and fortress-thick walls.

The Navesink Light, which is also known locally as the Twin Lights, is easily visited. In fact, each year over 90,000 people visit the old lighthouse and its accompanying museum. The station site, now a New Jersey state park, offers a commanding view of the Atlantic Ocean and the nearby coast. It is a good place to come, as many do, when the crowded steel and concrete canyons of the big city have grown wearisome and the soul feels the simple need for open skies, fresh sea breezes, and distant, unbroken horizons.

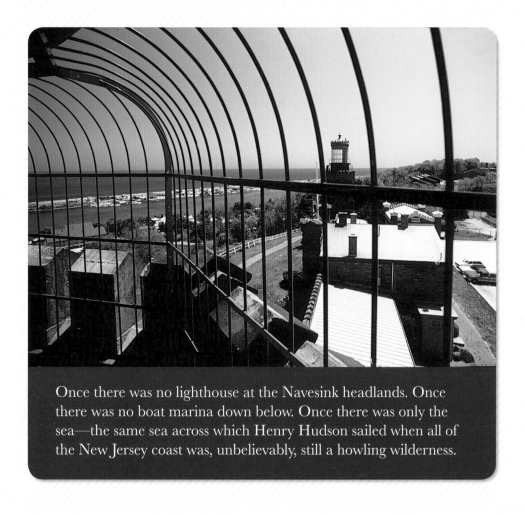

Once there was no lighthouse at the Navesink headlands. Once there was no boat marina down below. Once there was only the sea—the same sea across which Henry Hudson sailed when all of the New Jersey coast was, unbelievably, still a howling wilderness.

Tulips and dandelions blossom on the immaculately landscaped grounds surrounding the Navesink Light in New Jersey.

Minot's Ledge

Massachusetts (1850)

On such a tranquil day, it is hard to comprehend the terrible storm that once swept two lighthouse keepers at Minot's Ledge to an early death. The sea has her moods, both foul and fair.

ABOUT 20 MILES SOUTHEAST of Boston is the small community of Cohasset, which takes its name from a Native American word. Just offshore from Cohasset is one of the most dangerous shoals along the Massachusetts coast—the dreaded Minot's Ledge.

Part of the problem with Minot's Ledge is its specific location. In the past, vessels sailing toward Boston Harbor traditionally took their directions from the light at Provincetown and continued west on a steady compass heading. If, however, the ship was just a little bit off as it approached the coast south of Boston, drifting slightly to the south, it would unexpectedly ground and splinter timber on Minot's Ledge. If caught in a storm, if there were not enough lifeboats, or if the ship simply broke apart too quickly, a catastrophe and loss of life would ensue. Records show that literally dozens of ships wrecked on Minot's Ledge.

For decades, the construction of a lighthouse on the open sea at Minot's Ledge was considered an engineering impossibility. In 1847, however, a young engineer developed a plan to drive pilings deep into the bottom of the sea. On those pilings an iron structure would rise, and on top of that would be a small keeper's cabin and a tower holding a lantern. It was thought that the open-framed tower would offer less resistance to the wind and would thus be better able to survive a strong storm or hurricane.

The Minot's Ledge tower entered service on January 1, 1850, after a construction period of three years. The first keeper assigned to the tower, a man named Isaac Dunham, did not trust its safety and resigned after less than a year. His prudence and concerns, however, proved to be both sound and prophetic. Fifteen months after it was finished, during a powerful spring storm in the spring of 1851, the Minot's Ledge tower toppled over into the waves. In it were two young assistant lighthouse keepers—Joseph Wilson and Joseph Antoine.

John Bennett, the chief lighthouse keeper, was on shore, and could only watch in horror as the tower began to lean and finally collapse into the sea. Bennett later reported that the two lighthouse keepers bravely kept the light going until they literally fell into the surf. He was said to have never forgotten the steady ringing of the station bell over the crashing of the waves. For days afterward, the personal possessions of the keepers washed ashore on the beaches of Cohasset. Finally, their storm-wracked bodies, too, were laid to rest on the sandy shore.

The light's one-four-three sequence can be seen from 10 nautical miles away.

Minot's Ledge, Massachusetts (1850)

Nine years later, a replacement lighthouse was raised on Minot's Ledge. The new 114-foot tower was made according to traditional methods—with heavy blocks of stone rather than forged rods of iron. The upgraded Minot's Ledge Light was engineered so that any storm-driven waves would force its foundation blocks to bond together more closely. Steel shafts, cement, and interlocking cut stones were placed in such a way as to make a veritable sea fortress. Within its mighty walls, the new lighthouse keepers could feel safe from any storm and keep up the tradition of service so proudly upheld by assistant keepers Wilson and Antoine. As with many lighthouses, there are rumors of ghosts in the tower, on the nearby seas, and on the surrounding beaches.

The Minot's Ledge tower remains to this day, a symbol of both human tragedy and technological accomplishment. Even though two men died on the first tower, which can be called one of the greatest mistakes in nautical engineering, the second replacement tower stands as one of the great engineering achievements among U.S. lighthouses in the 19th century. The tower still flashes a beacon in a one-four-three sequence, which signals "I LOVE YOU." It can be viewed from the beaches of nearby Cohasset or on boat tours. A good museum in Cohasset, established by the Cohasset Historical Society, contains information and artifacts of interest about the historic Minot's Ledge Light.

The 97-foot Minot's Ledge Light has weathered countless fall hurricanes and winter storms over the years. It is a striking testament to humankind's ability to solve seemingly insolvable problems.

The historic lighthouse keeper's residence at Minot's Ledge has been used as a meeting hall by local civic leaders for many years. It is listed in the National Register of Historic Places.

The Minot's Ledge Light was rehabbed in the 1980s. This structure includes stones from the tower built in 1860 as well as a replica of the light room.

Block Island Southeast

Rhode Island (1875)

The deeply eroded cliffs of Block Island may be beautiful in this scene, but they pose a very real threat to the stability of the headlands on which the Block Island Southeast Light is located.

FIFTEEN MILES BEYOND Montauk Point, the easternmost spot on Long Island, is Block Island, a 25-square-mile piece of windswept land to the east of Long Island Sound. Roughly midway between the lighthouses at Montauk Point, New York, and at Point Judith, Rhode Island, the Block Island Southeast Light has served as an essential navigational aid for ships cruising into Long Island Sound or proceeding south toward New York City. The local waters are known for their submerged reefs, strong currents, and—because of the close proximity to the North Atlantic—their quickly developing fog banks and storms. This lighthouse has directed more than one vessel away from danger and toward safety.

The Block Island Southeast Light is distinctive in the size of its oversized catwalk platform, as can be seen in this magnificent image.

Visitors to Block Island normally take the ferry, but air transport is also available. What they find on their arrival is a quiet, scenic landscape that is far less crowded and hectic than Martha's Vineyard or Nantucket Island just to the north. The lighthouse itself, with its thick octagonal tower and attached manorial living quarters, resembles something from a Stephen King novel and is said to be haunted. Built of weathered red brick, with a steep roof of green shingles and windows with white frames, the lighthouse keeper's home evokes both the architecture and the way of life of the bygone Victorian era.

The lighthouse occupies a rugged but soft cliff of clay and sand that is eroding from the effects of wind, rain, and gravity. During the early 1990s, the lighthouse was jacked up on hydraulic lifts and moved back from the brink by nearly the length of a football field, where it will, it is hoped, be safe for many decades to come.

Block Island Southeast, Rhode Island (1875)

Although the light was deactivated in 1990, the entire 2,000-ton building was moved back from the cliffs by 300 feet because of the threat erosion had on the building's stability.

The lamp could warn mariners coming into or out of the Long Island Sound from 20 nautical miles away.

The Block Island Southeast Light has remained largely untouched by renovations since it was built, except for the addition of modern plumbing and repairs of storm-related damage.

The fog horn that was signaled once every 30 seconds in times of low visibility.

American Shoals

Florida (1880)

MENTION THE FLORIDA KEYS and most people generally think of writer Ernest Hemingway and his famous after-hours haunts, such as Sloppy Joe's Bar and Grill. Others envision world-class fishing or nearly year-round warmth and sun. Once a hideout for notorious pirates, Key West later became a thriving port and fishing center—and then a renowned vacation spot—on the northern edge of the Caribbean. Through it all, the Key West region has been a perilous one for mariners,

Recreational boaters always enjoy a stop at the American Shoals Light, where they can tie off for awhile and rest or perhaps fish the surrounding waters.

with numerous coral reefs, submerged limestone formations, and sandy shoals. Most of these hazards are impossible to see until one is nearly upon them. It is also an area frequently beset, from July through October, by killer tropical storms borne on the Southern Atlantic.

The deadliest reef of them all is located near Looe Key, about 20 miles north of Key West. Although no one has ever been able to compile a complete historical list of the ships that have gone down on the reef, the list of major vessels over the centuries probably numbers at least into the dozens. In the early years, local authorities erected various day marks over the reef, including a tall, unlighted piling. These, however, often did not alert mariners of the reef's presence until it was too late. At night, there was no way of alerting ships of the nearby danger.

Once Florida attained statehood, the United States Government began to build lighthouses up and down its long coastline on the Atlantic Ocean and Gulf of Mexico in real earnest. In 1878 Congress finally appropriated the necessary funds ($75,000) to build a lighthouse on the reef near Looe Key.

The American Shoals Light was actually built in the shipyards of New Jersey, and then transported by freighter over a thousand miles to the Florida Keys. The 109-foot spider-like tower was built atop a massive subsurface platform of rock. It included an octagonal lighthouse keeper's residence, located about 40 feet above the water, and a first-order Fresnel lens. Like many government projects, the cost of the American Shoals Light kept increasing (the final bill was $125,000). The light was fully automated in 1963 and is still operational. The flashing light, which has undoubtedly saved countless ships and lives, is said to be visible for about 12 miles.

In 2016, 24 Cuban refugees took shelter on the American Shoals Light, but they were later taken into custody by American authorities. Of the 24 refugees, 4 of them were repatriated while the others were interned at Guantanamo Bay Naval Station.

Cape Neddick

Maine (1879)

Although visitors to the Cape Neddick Light are separated from the tower by a short expanse of saltwater, the best views are actually from the nearby coast.

THE CAPE NEDDICK LIGHT, also known as the "Nubble Light," sits on the Nubble, a small stony island off the extreme southern coast of Maine at York Beach. Although inaccessible to visitors, the Cape Neddick Light is one of the more scenic lighthouses in New England. It can be viewed from York Beach across a stretch of saltwater.

Built in 1879, the Cape Neddick Light arrived toward the end of the great lighthouse building boom that began in the early 1850s. The light consists of a white wood-frame residence, a steep, dark-red roof, a covered walkway, and a 40-foot-high cast-iron tower that emits a flashing red signal light. The clapboard keeper's house is in the classic Victorian style, right down to the meticulous details under the eaves of the roof. Two strong wooden outbuildings, set a short distance from the lighthouse among the barren rock, complete the picture. Each Christmas the locals cover the lighthouse and keeper's house with electric Christmas lights, presenting one of loveliest holiday spectacles in the country.

Historically, this lighthouse has been important in keeping vessels away from the rocky stretch of coast between Portsmouth, New Hampshire, to the south and Portland, Maine, to the north. The nearby communities of York and Kennebunkport have grown into some of the wealthiest enclaves in America. A photograph of the Cape Neddick Light was included in the 1977 *Voyager* spacecraft, serving as a representation of Earth's civilization.

Christmas is definitely the best season to view the light display at the Cape Neddick Light, a scene that is often depicted on postcards and in calendars.

Cape Neddick, Maine (1879)

The Cape Neddick Light's first day of service was July 1, 1879. The tower was originally painted red but was repainted white when the keeper's house was built in 1902.

A photo from the early 1900s. The light was eventually automated much later in 1987.

Ponce de Leon Inlet

Florida (1887)

The Ponce de Leon Inlet Light is named for one of the first explorers of the American Southeast. After accompanying Columbus on his second voyage in 1493, Ponce de Leon later returned and searched for the mythical Fountain of Youth.

Ponce de Leon Inlet, Florida (1887)

SOUTH OF HISTORIC Saint Augustine, Florida, and north of Cape Canaveral, there is a long, low stretch of sub-tropical forests and open, sandy beaches. Although the coast appears cleaner from a navigational standpoint than the rocky coast of, for example, New England, maritime hazards still abound. For one thing, numerous offshore sand shoals, or sandbars, pose a real threat to vessels as they ply the waters parallel to the coast. For another, the coast is so flat that few opportunities exist to construct lighthouses from any natural feature resembling a headland or promontory. For this reason, lighthouse construction came both late and light to the region, despite considerable maritime activity dating back to Colonial times.

Only the physically fit can climb all of this tower's stairs. The Ponce de Leon Inlet Light is 18 feet shorter than the Cape Hatteras Light, which is the tallest in the U.S.

It was during the administration of President Andrew Jackson, in 1834, that the first lighthouse was constructed at Mosquito Inlet, Florida (as it was then referred to). This primitive structure was located just south of what is today known as Daytona Beach. The Seminole Wars erupted shortly thereafter, and the first Ponce de Leon Inlet Light—called the Mosquito Inlet Light—fell into a state of disrepair. Within two years, erosion became so severe that the tower fell over into the sand dunes.

Half a century later, the United States Lighthouse Board returned to the area and busily set about planning a new lighthouse to mark the shallow waters near the inlet. After all, a nearly 100-mile stretch of coast between Saint Augustine and Cape Canaveral had only one other light. Funds for the construction of a permanent light at Ponce de Leon Inlet were appropriated by the United States Congress in 1882, and the tower was ultimately completed in 1887. The tower stood 175 feet above ground level and was fitted with a first-order Fresnel lens. Today, thousands of Florida vacationers flock to the inlet to see the historic light, which is the second tallest brick tower in the United States.

The lamp atop the Ponce de Leon Inlet Light can be seen from 20 nautical miles away as it signals in groups of 6 flashes every 30 seconds.

Kerosene was the original fuel used for the lamp before it was replaced with an incandescent oil-vapor lamp in 1909. The lamp was electrified in 1933 and fitted with a first-order Fresnel lens.

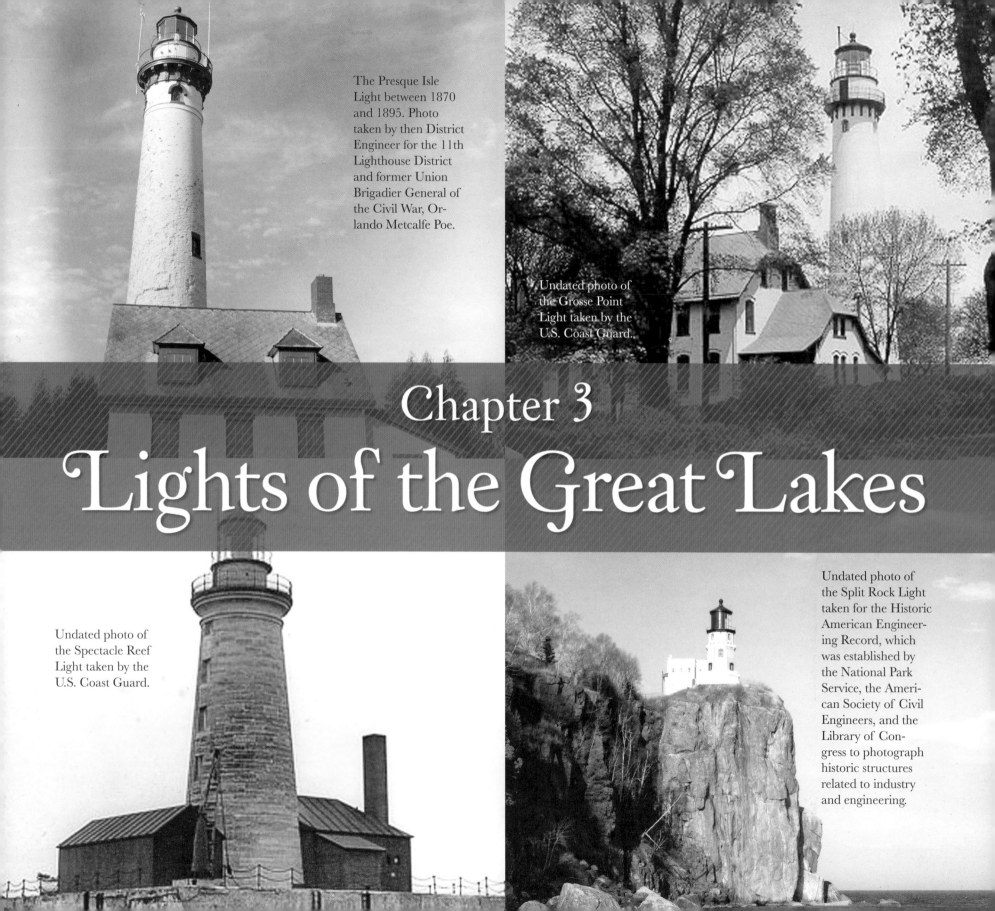

The Presque Isle Light between 1870 and 1895. Photo taken by then District Engineer for the 11th Lighthouse District and former Union Brigadier General of the Civil War, Orlando Metcalfe Poe.

Undated photo of the Grosse Point Light taken by the U.S. Coast Guard.

Chapter 3
Lights of the Great Lakes

Undated photo of the Spectacle Reef Light taken by the U.S. Coast Guard.

Undated photo of the Split Rock Light taken for the Historic American Engineering Record, which was established by the National Park Service, the American Society of Civil Engineers, and the Library of Congress to photograph historic structures related to industry and engineering.

Fort Niagara

New York (1782)

Surrounded by stately oak, elm, and maple trees, the gray stone lighthouse at Niagara rises eight stories above ground level.

Fort Niagara, New York (1782)

THE GREAT LAKES WERE BORN during the twilight of the last ice age, when the retreating continental ice shields in North America began to melt as the atmosphere warmed. The released water—billions of gallons of it—gathered in inland areas and slowly formed deep lakes. The results were the five Great Lakes—Superior, Huron, Michigan, Erie, and Ontario. These lakes are, for all intents and purposes, vast inland seas. They can produce waves, currents, ice flows, and fog banks every bit as treacherous as any found in saltwater. They also have shoals, reefs, and sandbars as perilous as those found in oceans. What's more, the lakes are known around the world for their savage, arctic-born winter storms.

The water connecting the first four lakes with Lake Ontario flows through one river: the Niagara River. Although the Niagara River is only about 40 miles long, more than 200,000 cubic feet of water surges through its rocky channel every second. Midway down the river are Niagara Falls, arguably the most spectacular waterfalls in the world. More than anything in the Great Lakes region, Niagara Falls provides a sense of how important water has been in shaping the landscape of the upper Midwest and southern Canada.

The outlet of the Niagara River in Lake Ontario has, since the beginning of human history in the area, been a natural place for people to gather. For one thing, because of the rich nutrients and oxygen in the churning river water, the fishing has always been good. For another, it was a convenient place to portage and trade. When the British entered the area in

Africa may have Victoria Falls, and South America may have Angel Falls, but North America has Niagara Falls, among the greatest spectacles of nature on the planet. The falls are located on the Niagara River, just a short distance from the Fort Niagara Light.

the late 18th century, they saw both of these things. They also brought with them their tradition of lighthouse building. For these reasons, they built a stone light tower atop Fort Niagara, which they had seized after the end of the French and Indian War.

After the American Revolution, the United States Government took control of Fort Niagara. Unfortunately, in the lean years following independence, the federal government could not afford to maintain the lighthouse, and the site fell into a state of disrepair. During the administration of Thomas Jefferson in the early 19th century, the original Fort Niagara Light was taken down.

About 20 years later, in 1823, a wooden tower was raised at the fort. The light, which stood above the mess house, was meant to mark the entrance to the Niagara River. During the 1820s, as the Erie Canal began to transport increasing amounts of goods west, freight trade on Lake Ontario steadily diminished. As a result, the Fort Niagara Light was relegated to a less important role in the region. This was particularly the case after Canada built the Welland Canal, which funneled more and more trade to Buffalo, bypassing the Niagara River altogether.

This cannon on the old military base of Fort Niagara bears testament to the fact that the northern coast of New York was not always a peaceful landscape. It was, at one point, contested ground.

Fort Niagara, New York (1782)

In 1872 a 50-foot octagonal stone tower was raised on the coast near the fort, and the old tower at Fort Niagara was decommissioned. Several more feet were added to the stately gray tower in 1900, which enabled its light to be seen more than 20 miles out on Lake Ontario. Although no longer an active station, the Fort Niagara Light was active until 1993. The lighthouse has gradually become the centerpiece of Old Fort Niagara, where visitors can enjoy summer military reenactments and various exhibits relating to the rich human and natural history of the Lake Ontario region. Old Fort Niagara is a half-hour drive north from Buffalo, New York.

Much has changed about the world since the present stone lighthouse at Fort Niagara was first raised in 1872 (four years before the Battle of the Little Bighorn!), but the need for reliable aids to maritime navigation still remains.

The year 2000 marked the 100th anniversary of the tower's last major renovation. The brick tower is sturdy enough to last another century.

Buffalo

New York (1818)

"Red at night, sailors delight. Red at morning, sailors take warning." The old adage seems particularly appropriate in this view of the lighthouse in Buffalo, with the evening sky as warm and red as a glass of wine.

Buffalo, New York (1818)

SINCE ITS EARLIEST DAYS, Buffalo, New York, has been important geographically and commercially because of its proximity to the natural resources of Canada and its strategic position on the Niagara River, the major waterway between Lake Erie and Lake Ontario. The site now occupied by Buffalo was a natural place for a regional political and business center to develop. Although a lighthouse was planned for Buffalo as early as 1805, it was not built until 1818, well after the War of 1812. That is probably just as well, since the war saw British troops set fire to parts of Buffalo, and a lighthouse would have been a prime target.

This was the first lighthouse to be built on Lake Erie. The creation of the Erie Canal in 1825 hastened the growth in importance of Buffalo as a vital commercial port and underscored the necessity of placing a good lighthouse on the southeastern shore of Lake Erie.

A new tower, raised in Buffalo in 1833, was constructed in the popular octagonal shape with masonry blocks made from locally quarried limestone. It was located at the end of a pier nearly 500 yards long, and it stood 68 feet above the waters of Lake Erie. The tower alerted vessels to the fact that they were approaching both the port of Buffalo and the Niagara River, which shortly turns into Niagara Falls. This tower was replaced in 1872 by a new lighthouse with a stronger light and a longer pier of 4,000 feet—nearly one mile!

Lovers of architecture admire the details of the Buffalo Light, one of the handsomest towers on the Great Lakes.

Tourists today actually visit the older lighthouse, which is on the grounds of the Coast Guard station near the downtown convention center. Although the original tower was derelict for nearly a century, Buffalo citizens began restoring it during the 1960s, and it has since become one of the most popular sites in the city.

The Buffalo Light was never automated like many other lighthouses built in the same era. It was deactivated in 1914.

A photo from an unknown photographer circa 1859. This structure replaced the original light built on this location in 1818, and it is the oldest lighthouse in Buffalo that is still standing in its original location.

Presque Isle

Pennsylvania (1819)

The light was automated in 1962 and can be seen from 16 nautical miles away.

ALTHOUGH MOST PEOPLE probably don't think of Pennsylvania as a state requiring a lighthouse, the truth is that the Keystone State has a significant corner of coastline along Lake Erie. In fact, on the long windswept coast between Cleveland, Ohio, and Buffalo, New York, the only other major port town on the lake is Erie, Pennsylvania.

Presque Isle, which in French means "almost an island," is actually a long peninsula that curves from the south like a protective arm around the port of Erie. In 1813 Presque Isle witnessed a unique moment in American history, for it was in the adjacent port of Erie that Oliver Hazard Perry raised a fleet of U.S. battleships that went on to beat the British in the Battle of Lake Erie. It was during this battle that the intrepid Perry uttered the famous words: "We have met the enemy and they are ours."

The Presque Isle Light was added to the National Register of Historic Places in 1983.

The original Presque Isle Light, which was actually located on the sand beach across from Presque Isle, was raised in 1819. The first tower was replaced in 1867. Circular in shape, it was built of sizable blocks of native sandstone. This light operated through much of the 19th century. Its replacement, the current Presque Isle Light, was built at the end of seven-mile-long Presque Isle in 1873. The 68-foot tower was outfitted with a Fresnel lens and was given a fixed white light. All those approaching the port of Erie, Pennsylvania, can count on this light.

A keeper's residence was built next to the tower, and a 1.5-mile road was created to connect the lighthouse with a boathouse in Misery Bay. Folks today take the leisurely stroll down Sidewalk Trail.

Visitors can find the Presque Isle Light located, conveniently enough, in the Presque Isle State Park. The earlier Presque Isle Light, also sometimes called the Erie Land Lighthouse, is also still standing. Both are surrounded by magnificent native hardwood trees.

Despite the popularity of the beacon today, the Presque Isle Light was once described as "the loneliest place on earth" by Charles Waldo, the first keeper of the station. On July 12, 1873, he wrote: "This is a new station and a light will be exhibited for the first time tonight." Later, he added a sad addendum: "There was one visitor."

A photo of the original 40-foot tower at the second location of the Presque Isle Light in 1885.

Marblehead

Ohio (1821)

The oldest active lighthouse on the Great Lakes, Marblehead Light is located on the headland of a peninsula that projects into the southwest corner of Lake Erie.

LIGHTHOUSES ON THE Great Lakes tended from the beginning to be built where population and business were growing quickly. There was a need both to warn of hazards and to signal mariners that they were approaching important ports of trade. Such is the case with Marblehead Light, the oldest active lighthouse in the Great Lakes region. The lighthouse was constructed to mark the entrance to Sandusky Bay on Lake Erie in northern Ohio. Located on Marblehead peninsula just to the north of Sandusky, the light directed navigators into the mouth of Sandusky Bay, which could sometimes be missed along the low, nondescript Ohio coastline.

The lighthouse at Marblehead, then called the Sandusky Bay Light, was built in 1821. It consisted of a sturdy 55-foot conical tower located on a rock platform just a few feet from the water's edge. Painted white, with red trim around the catwalk and on the roof of the light room, the lighthouse produced a strong light that could be counted on to show vessels the way into the bay. Eventually a Fresnel lens replaced the original array of 13 lamps and reflectors, and the tower was also raised some ten feet to increase its visibility out on the lake.

With the assistance of sensors and timers, the United States Coast Guard still operates a beacon at Marblehead Light, which is equipped with a fourth-order Fresnel lens.

One of the most interesting stories relating to the Marblehead Light occurred during the Civil War. A group of southern sympathizers tried in vain to liberate some of their compatriots from a Union prisoner-of-war camp on Johnson's Island, in Sandusky Bay just south of the Marblehead Light. As part of their failed effort, they stole a passenger vessel. The sight of a large Union ship guarding the island, however, scared them away. Instead of landing on Johnson's Island to liberate its prisoners, the Confederate partisans steamed across Lake Erie to Canada, where they sank their commandeered boat.

The lighthouse at Marblehead remains an active station, and the site has become an Ohio state park.

Marblehead, Ohio (1821)

Before the Marblehead Light was automated in 1958, it had been tended to by 15 keepers—two of which were women. The first keeper was a Revolutionary War veteran, Benajah Wolcott, who was one of the first residents to live in the Sandusky region.

The original light fixture found in the Marblehead Light was a set of 13 whale-oil lamps that would reflect off several tiny metal plates to project across the lake.

The original whale-oil lamps were replaced by a kerosene lamp and a Fresnel lens in 1858.

The lighthouse at Michigan City faithfully guided mariners for more than a hundred years, from the pre-Civil War era to the dawn of the modern age.

Michigan City

Indiana (1837)

MENTION THE STATE of Indiana and most people are not likely to think of lighthouses. The Hoosier State, located between the Prairie State of Illinois and the Buckeye State of Ohio, is known more for its flat cornfields and productive cattle-field lots than for large tracts of deep water requiring coastal navigational aids. However, Indiana does have an important 30-mile coastline on the southern border of Lake Michigan, between the vital regional trading center of Chicago and the various open-water ports along the western coast of Michigan.

It is along this northern Indiana coast of sand dunes and forested hilltops on Lake Michigan that the current Michigan City Light was raised in 1858 (the original light had been built in 1837). The Michigan City Light was designed to serve the busy harbor at Michigan City, an active shipping port for goods produced in central and northern Indiana. The lighthouse rises from a three-story keeper's residence. The tower emerges directly from the pitched red roof, giving the building the quaint appearance of some eccentric cottage in the English countryside that has decided to sprout a lighthouse from its roof. Although the Michigan City Light was decommissioned in 1904, local lighthouse aficionados have maintained the site as a museum.

Today the Michigan City Light can be found just a few miles east of the popular Indiana Dunes National Seashore, where hundreds of thousands of visitors throng every summer to enjoy the dunes, one of the Midwest's greatest attractions. The Michigan City Light is easily accessible either from South Bend, Indiana, which is located about 30 miles to the east, or from Chicago, which is about 60 miles to the west. The original Fresnel lens is housed in the lighthouse museum.

The lighthouse keeper at Michigan City had only to climb up the stairs through the roof of his home to check on the status of the light.

Selkirk

New York (1838)

With its weathered gray and black rock, pitched slate-covered roof, and short, red tower, the Selkirk Light resembles something from a Norman Rockwell painting.

AS WE HAVE SEEN, the Great Lakes present the same dangers and challenges to recreational and commercial vessels as do the saltwater coasts. Navigators must be aware of partially exposed reefs, sandy and muddy shoals, submerged rock formations, shifting channels, and prominent headlands. In bad weather, all of these hazards become even more perilous. Such is particularly the case on Lake Ontario, known for its late-fall and early-spring storms and its sudden summer squalls. In winter months, the weather can be even worse. The eastern lake region is frequently beset by bad weather, and that weather often develops quickly as fronts and storm lines meet and clash over the water. At such times, mariners can be in real trouble if they cannot identify hazards that exist along the coast.

It is no surprise, then, that one of the earliest lighthouses in the Great Lakes region was built at Selkirk Point near Pulaski, New York, on Lake Ontario. This unique lighthouse, near the mouth of the Salmon River, guards the lonely coast between Mexico Bay to the west and Stony Point to the north. The Selkirk Light is one of the prettiest small lighthouses in the Great Lakes area. It was built in 1838 of native fieldstone and secured with white cement, which produced a compact two-story structure resembling a sturdy New England farmhouse. This is a lighthouse that would fit well in a Norman Rockwell or Andrew Wyeth painting. It has the same spare austere quality in its construction and in its physical appearance. The light tower rises from the lakeshore side of the roof. The original light involved the old reflector and oil-lamp system, but the tower eventually was fitted with a Fresnel lens.

The Selkirk Light remained active from 1838 until the eve of the Civil War, when, like many lighthouses both in the Great Lakes region and elsewhere on coastlines across the country, it was shut down because of a lack of funds. It was not used again until 1989, when its light was restored for the benefit of the myriad amateur mariners and fisher folk who arrive at the lake every summer. The light is privately owned and is available to the public for short-term rentals. Local lore has it that the lighthouse is actively haunted.

The lighthouse's funding disappeared along with the plans to build a canal to connect Salmon River and Lake Oneida to the Erie Canal. With no canal in the area to keep the shipping industry alive, the lighthouse wasn't necessary.

Grand Haven

Michigan (1839)

QUITE OFTEN THE lighthouses on the west coast of Michigan, as well as elsewhere on the Great Lakes, are found near the outlets of major rivers. The reason is that human settlements naturally took hold in those areas, where resources were plentiful, traders could conveniently meet, and commercial fishing was reliable. As waterborne trade developed, a need arose to provide navigational aids in the form of lighthouses. These lights were generally more in the form of harbor lighting aids than primary coastal beacons.

Such was the case with the Grand Haven Lights, set near the entrance to one of Michigan's better harbors, the Grand Haven Harbor. Not only is the sizeable harbor a natural anchorage for lake vessels requiring a deep port, but it is also fed by the waters of the Grand River. The river runs through Grand Rapids, an important inland center of commerce 35 miles to the east.

The present lights at Grand Haven replaced two earlier towers (1839 and 1855) that had been on the mainland. The Grand Haven south pier inner light and the Grand Haven south pier head light are separated by a distance of over 200 feet. The first (south pier inner light, built in 1905) was a 50-foot cylindrical tower made of steel plates. On the top was a

Besides the two lighthouses at Grand Haven, dozens of lamps all along the pier illuminate the shoreline.

small light room. The second is a short, box-like structure that had served as the fog signal building. This second structure was moved to the pier extension in 1905 and had a lens room added. Subsequently, both towers were covered with iron plates to offer them better resistance against ice and waves.

Visitors to the town of Grand Haven can walk on the pier and view the two lighthouses up close. These unusual twin pier lights are among the most interesting and historic in the Great Lakes region.

The pier at Grand Haven is one of the longest in the Great Lakes region. Positioned as it is far from shore, the lighthouse is visible far out on the lake, warning mariners of shallow waters close to shore.

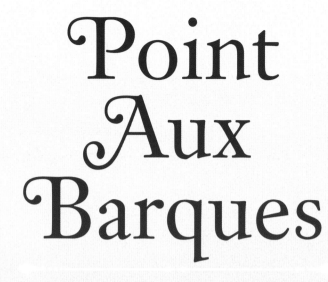

Point Aux Barques

Michigan (1848)

Calm waters surround the lighthouse at Point Aux Barques, near the head of Saginaw Bay on Lake Huron. For over a century and a half, the Barques Light has prevented calamities from occurring on the nearby waters.

SAGINAW BAY IS A LARGE, shovel-shaped projection of Lake Huron that cuts deep into eastern Michigan. Because its waters are so shallow—often less than eight feet—it represents a considerable hazard to large lake-traveling boats with any sort of draft. For this reason, the U.S. Lighthouse Board in the 1840s commissioned the building of a lighthouse on the prominent headland where Saginaw Bay meets Lake Huron.

This headland was historically called Pointe Aux Barques ("Point of Little Boats") by the French *coureur-du-bois* ("runners of the woods"). It was at this location that the early French fur-trappers, who traveled by birch-bark canoe, gathered each spring to trade winter furs and take on fresh supplies. Today, the nearby town of Port Austin is still a busy shipping port, although furs have been replaced by timber, fish, coal, copper, and other natural resources of the north country.

After the first lighthouse proved less than desired, a second was built in 1857 that still stands today. The current tower, painted white to increase visibility, is conical in shape and stands just under 90 feet. Although the lighthouse keeper is long gone, replaced by automated timers and sensors, the Pointe Aux Barques Light still shines with a flashing white light that is visible nearly 20 miles out on Lake Huron. The original Fresnel lens is on display at the Grice Museum in Port Austin.

The Pointe Aux Barques Light is particularly beautiful in the fall. The surrounding oaks and maples turn brilliant shades of red and orange, the Michigan sky is a clear autumnal blue, and the waters of Lake Huron are touched here and there with whitecaps.

The white conical lighthouse at Point Aux Barques is among the most photographed lighthouses in the Great Lakes region. Thousands of tourists flock to the popular site every summer.

Holland Harbor
Michigan (1872)

Looking more like some sort of overgrown firehouse than a lighthouse, the Holland Harbor Light is a favorite of artists and photographers.

THE OUTER COAST of Michigan on Lake Michigan is over 300 miles long. It runs from the open country near the Indiana state line to the heavily forested and rugged shoreline near the Straits of Mackinac in the far north. It is for the most part a low, flat coastline, with periodic river and stream outlets. Sandy dune fields and clay bluffs rise now and then, and the land becomes more hilly to the north. Major hazards to mariners include sandy shallows, mud shoals, and offshore rock reefs. Nowhere along that coast is there a more important lighthouse than at Holland Harbor, also known as Black Lake.

Traditionally, commerce on Lake Michigan proceeded north and south in shipping lanes that were parallel to the coast. Thus, from the beginning, a system of lighthouses was built that prominently lit the coasts. The light at Holland Harbor was, and still is, an integral part of that important coastal navigational system. The lighthouse is about a third of the way up the Michigan coast and roughly due west from the inland community of Lansing. Ships headed south toward Gary and Chicago depend on the light at Holland Harbor to keep them on course—and well away from the Michigan coast and its dangerous submerged limestone reefs and sandbars.

Known as "Big Red," the current Holland Harbor Lighthouse was built during the Great Depression, in 1936, and sits at the end of a long pier. It is a sizable structure, rising five stories above the pier, and is painted cherry red to increase its visibility to mariners out on the lake. The building is also covered with steel plates to protect it from the ice and waves of the winter storms. The Holland Harbor Light is an active station, and at night its light is visible up to 14 miles.

It's a glorious day at the Holland Harbor Light, with the Stars and Stripes flapping proudly in a steady wind. The red-painted lighthouse is quite unusual, and stands out brightly against the green foliage along the shore.

Holland Harbor, Michigan (1872)

Big Red also provides guidance through the channel to enter Lake Macatawa from Lake Michigan.

The present structure at this site was built in 1907.

Although it was slated to be decommissioned in 1970, residents of Holland petitioned the move and were able to preserve the lighthouse and its operations. It was automated that same year.

The original Fresnel lens used was a fourth-order lens, which is the median size lens Fresnel made for use in lighthouses.

No doubt about it, the keeper's residence at the Grosse Point Light is the nicest in the country. It seems more like the home of a university president than a lighthouse keeper.

Grosse Point

Illinois (1873)

THE "WINDY CITY" OF Chicago emerged as the unofficial capital of the Great Lakes region as early as the 1820s, when the Erie Canal brought a wave of eastern immigrants to the area. Later, the construction of railroads connecting Chicago with the east and west coasts resulted in another boom. Although a third of the city burned down in 1871, it quickly rebounded during the decade that followed with traditional Midwestern vigor. Through it all, Chicago continued to grow as the single most important port on Lake Michigan, where huge amounts of grain, meat, and other manufactured products were transported through a vast system of shipping routes. At the turn of the 20th century (1900), Chicago was nearly as busy a port as New York and San Francisco.

Because of its commercial importance as a center for regional trade, a series of lighthouses were built along the coasts east and west of Chicago. Of these, the loveliest of all was the Grosse Point Light, which was raised in Evanston in 1873. The conical white tower, which stood at 113 feet, helped navigators of Chicago-bound ships find their way along the coast. It showed a powerful second-order Fresnel lens. On the station grounds down below, the Italianate lighthouse keeper's house was a true thing of

The Grosse Point Light is among the top historical spots to visit in the Windy City. Few lighthouses in the system are as accessible to the traveler or the urban resident.

beauty, and eventually became one of the most sought-after duty stations in the Lighthouse Service. (Compare this to lighthouse duty at, say, Cape Flattery, a lonely windswept island far from good schools, clinics, or restaurants.)

Although the property of Grosse Point Light was turned over to the city of Evanston in 1935, it continues to be maintained as both an active, private aid to navigation and as the centerpiece of the city's Lighthouse Park.

The light was decommissioned in 1941, but was reignited in 1945 as a secondary navigational aid. It is operated by the Lighthouse Park District of Evanston, Illinois.

Grosse Point, Illinois (1873)

The Grosse Point Light is projected by a second-order Fresnel lens, which is the largest Fresnel lens found on the Great Lakes. There are 70 Fresnel lenses still in use in lighthouses through the U.S. today, 16 of which are on the Great Lakes.

It is rumored that the site in which the Grosse Point Light is located is also where Father Jacques Marquette landed to meet with local Illiniwek during his 1674 exploration of Lake Michigan's west coast. Father Jacques Marquette founded the first European settlement in Michigan, Sault Ste. Marie, and was also the first European to winter in what would later become the city of Chicago.

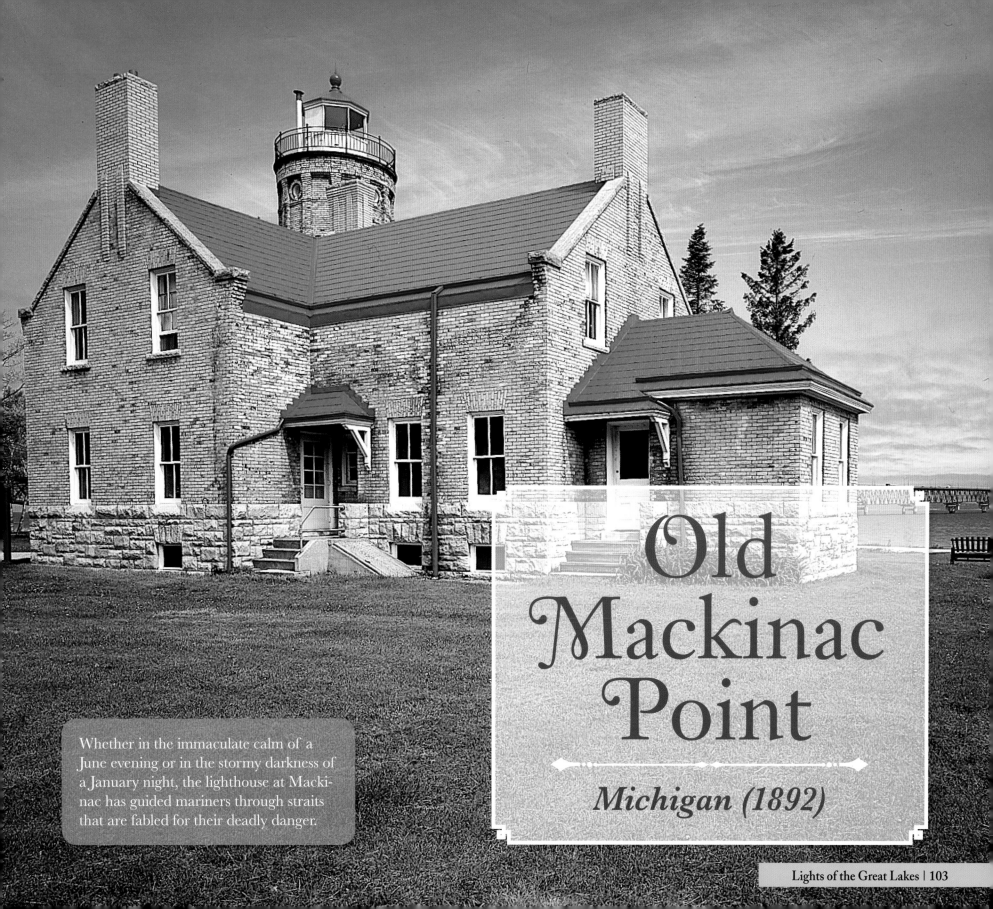

Old Mackinac Point

Michigan (1892)

Whether in the immaculate calm of a June evening or in the stormy darkness of a January night, the lighthouse at Mackinac has guided mariners through straits that are fabled for their deadly danger.

Old Mackinac Point, Michigan (1892)

NORTHERN MICHIGAN and the Upper Peninsula offer some of the most fascinating country in the Great Lakes region, both in terms of human history and natural history. This is the historic homeland of the Chippewa Indians, who paddled over three lakes—Huron, Superior, and Michigan—in birch-bark canoes. In lieu of lighthouses, the Chippewa used wood-burning fires along the lake shorelines. As early as 1616, French explorers came into this wild, little known area. These adventurers included famed Jesuit priest Father Marquette. The Father Marquette National Memorial and Museum is located on the Upper Peninsula just above the Straits of Mackinac.

Since the beginning of water-borne travel on the Great Lakes, the Straits of Mackinac have posed special challenges to

mariners. At the straits, the waters of Lake Michigan and Lake Huron mix and intermingle through a narrow, six-mile-wide channel. The result is sometimes dangerous currents that form in an area known for its submerged rock reefs, quick-developing fogs, and violent storms. In 1892 the first substantial lighthouse was built at Mackinaw City on the south shores of the Straits of Mackinac. The 40-foot light tower, built of gray masonry, rose near a two-story lighthouse keeper's residence. The sturdy stone residence, with a pitched red roof and three chimneys (winters are cold in the far north), looked more like a small castle than a lighthouse keeper's home.

The construction of the bridge over the Mackinac Straits in 1957 made the Old Mackinac Point Light unnecessary, and so it was decommissioned. Today the area is part of a modest city park overlooking the historic Straits of Mackinac.

Rising with castle-like walls and turrets, the massive brick lighthouse in Mackinaw City, Michigan, is one of the more solid and imposing in the Great Lakes region. The shores of nearby Canada are easily seen across the cold, blue waters.

Before European settlers arrived in this area, the waters of the Straits of Mackinac were treacherous for those traveling between Michigan's Upper and Lower Peninsulas.

The Mackinac Bridge, seen beyond the tower in this picture, made the lighthouse obsolete because the lights on the bridge were a much more effective aid to navigation.

The fourth-order Fresnel lens was visible from 14 nautical miles away.

Spectacle Reef

Michigan (1874)

The tower stands 95 feet tall and is located on a submerged reef about 11 miles east of the Straits of Lake Huron. It was featured on a series of stamps nominated by the Lighthouse Preservation Society and issued by the United States Postal Service in 1995.

WHEN ERNEST HEMINGWAY was a boy growing up in the suburbs of Chicago, his father, a prominent physician, took his family to northern Michigan every summer. There they stayed at a small community by Lake Huron and spent their time hiking and fishing. Some of Hemingway's best short stories were inspired by those family vacations in the north country near the Canadian border. While Hemingway was still small, there had already been a major lighthouse at nearby Spectacle Reef for over 30 years. The Spectacle Reef Light alerted Great Lakes mariners to an underwater reef that the U.S. Lighthouse Board once said was "probably more dreaded by navigators than any other danger now unmarked throughout the entire chain of lakes." The dangerous reef was a submerged rock platform about ten miles east of Bois Blanc Island, located in Lake Huron just east of the Straits of Mackinac.

Lighthouse experts consider the Spectacle Reef Light, completed in 1874 at the then-considerable cost of $400,000, to be "one of the outstanding [engineering achievements] in the lighthouse service as a whole." In order to build the foundation for the lighthouse, an elaborate crib-like dam had to be built to allow workers a dry area in which to proceed with their task. Once the crib dam was raised and the lake water pumped out, the base of the lighthouse was carefully constructed with bolted, precut stones that were cemented together.

The circular lighthouse rose nearly eight stories above the level of the lake. Its powerful light warned mariners of the dangerous shallows, shoals, winter ice jams, and rock-encrusted reefs that lurked near the Huron narrows. Even today, in the 21st century, the beacon of the Spectacle Reef Light continues to light the darkness and alert vessels that death is as near as the rocks on which the lighthouse rests.

The light tower at Spectacle Reef projects a light with a maximum candle-power of 400,000. Over the past 120-plus years, this light has undoubtedly saved countless ships and lives.

Sand Island

Wisconsin (1881)

BEAR ISLAND. Rocky Island. Oak Island. Outer Island. Cat Island. Stockton Island. Sand Island. Together this loosely scattered family of 22 timbered, rocky islands in southern Lake Superior are known as the Apostle Islands. They are among the most cherished natural areas in all of northern Wisconsin. They were also, in days gone by, the site of six light stations that helped guide mariners through the often confusing archipelago and toward such busy nearby ports as Duluth. Among those light stations

A rainbow shines brightly over the beautiful lighthouse at Sand Island. With its thick walls of mortar and country rock, this is a lighthouse that will be standing a century from now.

were Raspberry Island Lighthouse (1863), Outer Island Lighthouse (1874), Devils Island Lighthouse (1891), Michigan Island Light (two towers, 1857 and 1880), and Sand Island Lighthouse.

Of these, none are more lovely than the Sand Island Lighthouse, which was constructed in 1881. The sandstone tower exhibits a light 52 feet above Lake Superior. It helps to guide vessels south along the busy shipping channel to Duluth, some 50 miles distant. Like all the other lighthouses on the Apostle Islands, the light is accessible by boat or ferry, and is open to the public. Together these handsome, well-built, freshwater lighthouses form the world's most extensive outdoor lighthouse museum.

Today the Apostle Islands are collectively protected as the Apostle Islands National Seashore (AINS). Managed by the National Park Service, the AINS protects a pristine vignette of the north country as it once was, and of how important lighthouses remain to recreational and commercial traffic on the Great Lakes. Just a one-hour drive from Duluth, the breathtaking islands offer visitors not only the chance to recreate in some beautiful wild country, but also the opportunity to view some of the finest lighthouses ever constructed on the Great Lakes.

On most days, the wind blows strongly along the wild coast near Sand Island. At such times, the historic lighthouse serves a vital service in alerting boaters to the presence of the headland.

Split Rock
Minnesota (1910)

NORTHERN MINNESOTA IS famous for its many beautiful places, such as the Boundary Waters Canoe Area, Voyageurs National Park, the Lake of the Woods country, and the rugged Lake Superior coastline. It is a vast, wild territory that, despite a few roads and towns, has changed little since the first French "runners of the woods" paddled their birch-bark canoes across the Great Lakes in the 17th century to begin exploring. It is a landscape in which nature rules supreme, a place where fall storms roar in from the Arctic every October and spring snows

Split Rock is a photographer's and artist's dream come true, especially in the fall, as can be observed here: bright, white clouds, red and orange foliage, weathered gray cliffs, and deep blue waters.

linger into May. It's where the black bears grow to legendary size and the grey wolves go running over the frozen lakes beneath the shimmering northern lights. Residents and visitors alike have observed that nowhere in all that fabled north country is there a site more lovely than Split Rock Light.

The Minnesota coastline along western Lake Superior runs from the Thunder Bay region on the north to the port city of Duluth on the south. For the most part, it is a heavily forested

As can be observed in this aerial shot, Split Rock is one of the most spectacular lighthouses in North America, with the nearly vertical cliffs of rock and the colorful hardwood forests on the adjoining terrain.

coastline, with rolling hills, steep ridges, and frequent rock outcrops. Although the lake waters are deep on the northern coast—ranging to 129 feet near Grand Portage, Minnesota, and Isle Royale, Michigan—they become quite shallow near the lake's western coast in Minnesota—down to 20 feet or less. For this reason, a series of coastal lights was needed to warn vessels. Since the beginning of its modern exploration, Lake Superior—the largest of the Great Lakes at more than 350 miles long and covering 31,700 square miles—has seen a great amount of commercial traffic. Products such as timber, fish, wheat, corn, coal, iron ore, copper, and automobiles have traveled along its shipping lanes. The need for coastal lighthouses became more pressing over time.

Split Rock, Minnesota (1910)

Like many lighthouses on the Great Lakes, the lighthouse at Split Rock was built fairly late—1910. Part of the impetus for placing a station at Split Rock was that the schooner-barge *Madeira* was lost and the steamer *William Edenborn*, which had been towing the *Madeira*, wrecked near the site during a terrible storm in November 1905—a storm that sank or damaged more than 30 other boats on Lake Superior as well.

The Split Rock Light was constructed on about seven acres of land roughly 20 miles northeast of Two Harbors, where a lighthouse had been established in 1892 to guide iron-ore freighters and other commercial vessels in the busy Minnesota channel. The actual task of raising the Split Rock Lighthouse was quite difficult, as building materials had to be transported by boat from Duluth, nearly 50 miles to the south, and then hauled up to the top of the 120-foot Split Rock cliffs. The tower and residence cost over $70,000 in materials and labor, the equivalent of several million dollars in today's currency.

One of the most interesting aspects of the Split Rock Lighthouse was the clockwork machinery that turned the giant Fresnel lens and caused the light to flash. Its inner workings resembled those of a grandfather clock. A cable attached the turntable and various gears to a 200-pound weight, which was slowly lowered down the center of the lighthouse tower. Every few hours, the lighthouse keeper had to use a hand device to winch the weight back to the ceiling so that it could continue to move the heavy lens apparatus. The lens produced a beam that could be seen from 20 miles out on the lake, and remains in place to this day. There is no telling how many lives it saved during its 59 years of service.

A close-up photograph reveals the classic octagonal shape of the lighthouse at Split Rock, a design that has helped it weather the punishing winds of winter storms.

Today nearly 200,000 visitors flock to the lighthouse at Split Rock annually. Of all the lighthouses on the Great Lakes, it is perhaps the most scenic. It is certainly one of the most easily viewed, located only about an hour's drive north from Duluth. The octagonal, yellow-bricked tower is 54 feet high and offers a commanding view of the shoreline and of Lake Superior. Decommissioned in 1969, the Split Rock Light is no longer active, but it does house an important lighthouse and maritime museum.

This rugged, wintry scene evokes all of the desolation of the Great Lakes coast at that time of year. Who knows how many ships were saved over the decades by the faithful beacon of this lighthouse?

The lens and machinery at the core of the lighthouse are both simple and complex—simple in the physics of light, but complex in the mechanisms required to project them and rotate them faithfully for years at a time.

The Biloxi Light in 1901.

The Biloxi Light in 1906.

Chapter 4
Gulf Coast Lights

The Port Isabel Light in 1934.

The Sanibel Island Light in 1933.

Biloxi

Mississippi (1848)

The Biloxi Lighthouse owns the distinction of being the only lighthouse in the world that is currently—as a result of development—located in the median strip of a busy highway.

Biloxi, Mississippi (1848)

MANY OF THE LOCAL MYTHS that grow around America's lighthouses are just that—baseless legend and fable. A case in point is the Biloxi Light on the Gulf coast of Mississippi. In days gone by, one of the popular legends concerning the Biloxi Light was that it was painted with a coat of black paint following the assassination of President Abraham Lincoln in April 1865. What actually happened was that the lighthouse was originally painted black but was repainted white so as to make it stand out more against the background of live oak trees. And this occurred in 1867—two years after the president's death.

In a beautiful town with many wonderful coastal sites to visit—the Gulf Islands National Seashore, the Maritime Museum, the Marine Life Oceanarium—the Biloxi Light stands out. Located about six miles west of the entrance to Biloxi Harbor, the Biloxi Light was built in 1848. The 48-foot conical tower was made of cast iron, with an inner masonry wall of locally fired bricks. Mounted on top of the tower was a fourth-order Fresnel lens. One of the unusual facts about the Biloxi Light is that it has had two female lighthouse keepers. Maria Younghans, who held the job from 1867 through 1920, was succeeded by her daughter Miranda, who stayed until 1929.

The Gulf Coast is famous for its powerful hurricanes, which are born of the warm waters held between the low, humid panhandle of Florida and the Mexican tropical coast. However, the Biloxi Light has withstood them all. Today the Biloxi Ligh is one of the easiest lighthouses to see in the United States. It is, believe it or not, located in the grassy median strip between the eastbound and westbound lanes of U.S. 90.

The locals love to adorn the Biloxi Light with red and green strands of electrical lights each year during the holidays. The rising solstice moon of December completes this seasonal picture.

The Biloxi Light is the only light that was kept for more years by female keepers than male keepers.

The Biloxi Light was reopened to the public for tours in 2010 after a $400,000 restoration project that took 14 months to complete.

The Biloxi Light has survived many deadly hurricanes over the decades, including Hurricane Katrina which caused waters to surge over nearly one-third of the tower's 64-foot height.

Point Bolivar

Texas (1852)

Last light breaks over the quiet maritime landscape surrounding the Point Bolivar Light. At other times of year, hurricanes turn this scene of tranquility into one of nature's fury.

THE GULF COAST OF Texas is quite similar to the Atlantic coast of North Carolina. In both cases, a series of offshore barrier islands, and an intervening bay-like waterway, offer substantial protection to the state's mainland. Scattered among the barrier islands are various passes and inlets through which maritime traffic can safely proceed. These entryways provide access to the intracoastal waterways as well as to a number of important ports, most notably Port Arthur, Galveston Bay, and Corpus Christi Bay. Galveston Bay is probably the most famous of the three. It is protected by the Bolivar Peninsula to the north and Galveston Island to the south.

It was on the Bolivar Peninsula that the Point Bolivar Light was raised in 1852, during the heyday of lighthouse construction along the Gulf and Atlantic coasts. After the Civil War, Galveston became the point through which many European immigrants entered the American Southwest. So it was only natural that an important lighthouse be raised and maintained at the entrance to Galveston Bay. The first tower at Point Bolivar was dismantled during the Civil War by Confederate troops, who used the tower's iron to make cannon balls and mini-balls for rifles. In 1873 a new 116-foot tower was raised at Point Bolivar. It was again constructed of cast-iron plates, but this time a brick lining was installed inside the tower.

The lighthouse at Point Bolivar gained national recognition in September 1900 and then again in August 1915, when a number of local residents sought refuge in the tower from deadly storms. During the 1900 hurricane, over 120 people found shelter in the tower—literally huddled in the tower's spiral stairway. During the 1915 hurricane, some 60 people sought refuge in the same place from the hurricane. It was the infamous September 1900 hurricane that killed over 8,000 of Galveston's 30,000 residents. In the years that followed the disastrous 1900 hurricane, the Army Corps of Engineers raised a ten-mile-long and 17-foot-tall sea wall around Galveston, which has since become Galveston's version of the Ocean City boardwalk.

The Point Bolivar Light was decommissioned during the 1930s. Although the lighthouse is not currently in operation and is on private land, it is a popular site for people visiting the greater Galveston area.

The Point Bolivar Light is now located on private property and is not open to the public.

Point Isabel

Texas (1853)

THE SOUTHEASTERN GULF coast of Texas is distinguished by a number of long, flat barrier islands, generally set about ten to 20 miles from the mainland. Perhaps the best known of these barrier islands are North Padre Island and South Padre Island, which are also the sites of a nearly 100-mile-long national seashore. Between these offshore islands and the Texas coast is the Laguna Madre, which is a vital part of the Intercoastal Waterway (a naturally protected channel along the Gulf in which ships may safely proceed during all sorts of inclement weather). It is on the extreme southern point of South Padre Island that the famous Point Isabel Light is found.

The cylindrical design was preferred by lighthouse builders in the mid-Atlantic and southern states, as was the case at Cape Hatteras, Cape Florida, Biloxi, and here, at Point Isabel.

The Point Isabel Light was built in 1852, during the mid-19th century heyday of lighthouse building. It was actually constructed on the site of an army camp that was commanded by General Zachary Taylor (the future American president) during the Mexican War. The lighthouse, which is only a few miles north of the Rio Grande River and the international border with Texas, was designed to alert passing ships both to the mouth of the Rio Grande and to the presence of Padre Island to the west.

The 57-foot conical light tower was briefly decommissioned in 1888—a result of diminishing sea traffic in the region as more growth and trade occurred further west along the Pacific Coast. However, it was subsequently relit in the 1890s. The light was permanently turned off in 1905, during the administration of President Theodore Roosevelt. The lighthouse is now the center of Point Isabel State Historical Park, a popular day trip from nearby Brownsville, Texas.

Much has changed about the world since the light at Point Isabel was first raised over a century ago. Today a busy highway, a bridge, and associated development have transformed the once wild and remote landscape.

Sabine Pass

Louisiana (1856)

The large side supports of the Sabine Pass tower hold it erect despite the softness of the marshy grounds that surround it. The lighthouse is a visible reminder of the ingenuity and persistence of the human spirit.

THE COAST OF LOUISIANA is quite different from coasts further to the east and west. It is a much more complicated coastline, particularly around the delta of the Mississippi River, with myriad bays, channels, islands, peninsulas, river outlets, forested bayous, and open swamps. In this region the word "pass" is used by locals to signify a natural, fairly deep water route through substantial shipping obstacles, such as shoals, sandbars, and low water. Two of the most important passes are near the furthest delta of the Mississippi River.

Although the early French settlers established primitive beacons as early as 1721, the first substantial lighthouses were not built until much later. One of those was the light at Sabine Pass, which is located near the mouth of the Sabine River at the border of Louisiana and Texas. This light was designed and built by Army engineer Danville Leadbetter. An 1836 graduate of the U.S. Military Academy, Leadbetter was, along with George Meade (both later active as officers in the Civil War), one of the most eminent of the early lighthouse builders. His design for the Sabine Pass Light included enormous buttresses to give the structure better support in the soft, perpetually wet ground. Although a fierce hurricane in 1886 swept away the keeper's house, the tower stood firm. That it still remains a century and a half after it was raised says something about the ability of its creator.

The Sabine Pass Lighthouse was active until 1952, when it was decommissioned. For many years it was located on private land, but in 2001 the property was given to the Cameron Preservation Alliance. A four-mile road has been built to the lighthouse, making it accessible by land for the first time. The Sabine Pass Lighthouse grounds are open, but the tower remains closed.

The Gulf Coast is home to some of the toughest plants in North America. This subspecies of the common prickly pear cactus deters everything from foraging deer to corrosive saltwater. Just beyond, the Sabine Pass Light is also built to last.

Sanibel Island

Florida (1884)

Lush hardwood forests and panoramic blue-green tropical waters surround the historic light at Sanibel Island. Some of the best seashell collecting in America is found on nearby beaches.

The lighthouse on Sanibel Island is sturdily built to withstand the strongest of hurricanes. Though winds in excess of 125 mph have swept over the island, the lighthouse endures.

SANIBEL ISLAND IS ONE of the jewels of the southern Florida coast. Located just a short drive from both Miami and Tampa, the island is known around the world for the exotic seashells found along its white beaches. The island also occupies a strategic maritime location with respect to the mouth of the Caloosahatchee River, which drains the swamps east of Lake Okeechobee, and popular Charlotte Harbor to the north. Offshore sandbars and shoals are of further concern to passing ships. For this reason, a navigational lighthouse has stood on the island for well over a century, since 1884.

The story of how there came to be a lighthouse on Sanibel Island begins, as is often the case, with a shipwreck. In this case the very ship carrying the raw iron to build the lighthouse sank a few miles offshore after becoming grounded on a shallow-water sandbar. Vessels from Key West steamed up the coast and were able to salvage some of the cargo. The lighthouse was completed during the summer of 1884.

The tower is unusual in construction. It rests, windmill fashion, within an interlocking iron framework that, in turn, is attached to concrete supports seated deep in the ground. A winding staircase leads over 120 steps to the lantern house, where a modern flashing light alerts passing ships.

The Sanibel Island Lighthouse is also located adjacent to the Darling National Wildlife Refuge—a 4,975-acre wetland sanctuary where over 700,000 visitors each year view herons, storks, bald eagles, ospreys, alligators, otters, sea turtles, and other wildlife. Visitors will find both the lighthouse and its wild environs a delightful place to visit during the winter, well after hurricane season and well before the hot midsummer months.

The Old Point Loma Light around 1934, 43 years after it was decommissioned.

An undated photo of the Cape Flattery Light, with USCG *Walnut* approaching Tatoosh Island for the light's scheduled maintenance.

Chapter 5
West Coast Lights

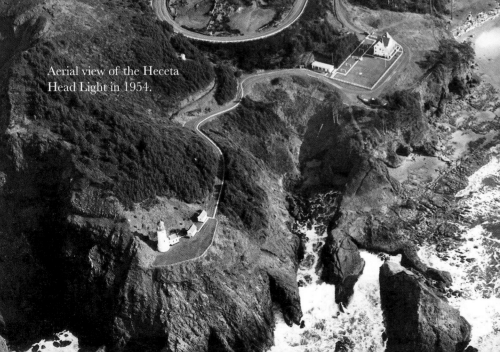

Aerial view of the Heceta Head Light in 1954.

A photo of the Admiralty Head Light in 1903.

Old Point Loma

California (1855)

Dusk comes early in the winter months to the California shores. When it did in past years, mariners were sure that the light at Old Point Loma would go on, serving as a faithful beacon to all those who journeyed at night.

Old Point Loma, California (1855)

This close-up view of the light room at the top of the Old Point Loma tower reveals the sturdy construction—heavy glass, riveted steel, contoured roof—that helped it withstand the winter storms.

JUST 12 MILES NORTH of the border with Mexico, on a sandy peninsula of land that curves like a protective arm around San Diego Harbor, is one of the most scenic lighthouses in the United States: Old Point Loma. Located on the grounds of Cabrillo National Monument, which commemorates the 1542 exploratory voyage of Juan Rodriguez Cabrillo, the lighthouse stands out prominently at the top of a salience overlooking San Diego.

Long before a lighthouse was built on the site, colonial Spanish regularly built signal fires on Point Loma, to help guide their vessels into the safety of San Diego Harbor. The present lighthouse was constructed in the years following California statehood (1851–1854). During that time Californians were anxious both to solidify an American presence on the historically Spanish coast and to facilitate active commercial passage into San Diego Harbor. Once authorized, the lighthouse was built with sandstone bricks chipped by masons from the local hillsides. The light tower was raised directly from the dark shingled roof of a compact, one-story structure, which was painted white to aid its visibility.

Because of the massive amounts of fog in the area, the light was extinguished in 1891 and a new light was built closer to the oceanfront.

Topped with a Fresnel lens, the Old Point Loma Light remained active until 1891, when it was closed. Although the lighthouse was nearly 500 feet above the level of the sea, it was also well up into the fog zone, and hence its light was frequently obscured. The lighthouse was eventually replaced with a skeleton tower at a lower, fog-free elevation. Visitors to the Old Point Loma Light today are able to spot migrating gray whales far out to sea. The view from Old Point Loma has been called one of the three best harbor views in the world.

The Old Point Loma Light is one of the first eight lighthouses built on the Pacific coast. It is only open to the public on August 25, the National Park Service's birthday, and November 15, the lighthouse's anniversary.

During WWII, the house was painted in a camouflage color-scheme and was used to signal naval ships into San Diego Harbor.

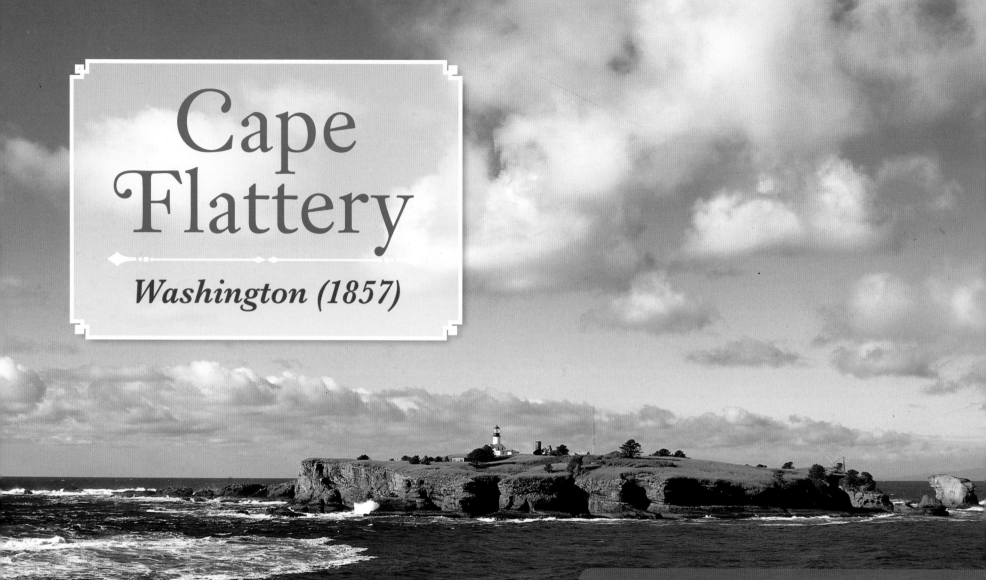

Cape Flattery

Washington (1857)

The remote lighthouse at Cape Flattery, northernmost in the lower 48 states, marks the stormy entrance to the Strait of Juan de Fuca west of Seattle.

CAPE FLATTERY WAS NAMED by Captain James Cook on his celebrated voyage of discovery around the world in 1778. Why Captain Cook named it "Cape Flattery" is not known. What is known is that few lighthouses are more isolated than this one. Located on a small island about 30 miles south of Vancouver Island, British Columbia, the Cape Flattery Light marks the southwestern entrance to the often stormy Strait of Juan de Fuca.

The Cape Flattery Light was established on Tatoosh Island in 1857, just one year after the first lighthouse in the Pacific Northwest was built at Cape Disappointment on the Columbia River. The installation at Cape Flattery included a modest residence in the popular Cape Cod style of the period, as well as a 65-foot stone tower that rose directly from the red, metal roof. Because the top of Tatoosh Island is itself 100 feet above the level of the sea, the light was quite high for the area—165 feet above the surf line.

Tatoosh Island was, at the time, a holy area for the Makah coastal Indians of northwestern Washington. The Makah came by canoe to the area to bury their dead. In fact, when one of the early government expeditions arrived on Tatoosh Island in the early 1850s, they found quite a few Makah Indians waiting for them. Although no hostilities broke out, subsequent expeditions spread a deadly smallpox epidemic among the Makah. Relations between the United States Government and the Makah Indians remained strained for years.

Twenty-six years after the Cape Flattery Lighthouse was built, in 1883, the federal government added a weather station to the island. Still later, a radio transmitter was built on the island. Eventually, more than 40 people lived around the lighthouse, and a small schoolhouse was even built there. Change came permanently, though, in 1966, when the weather station was decommissioned and most of the inhabitants were forced to leave. Although unmanned since 1977, the light, fog signal, and other navigational aids are still maintained by the Coast Guard, which flies in periodically with helicopters to check on the remote station. Unlike many other lighthouses built in the 1850s, the Cape Flattery Light remains an essential beacon to commerce.

The Cape Flattery Light was originally fitted with a first-order Fresnel lens which was replaced with a fourth order in 1932. A modern optic lens replaced the fourth-order lens in 1977.

St. George Reef

California (1892)

The St. George Reef Lighthouse often attracts seals, sea otters, and sea lions. At times, the din of their social barking and squealing drowns out the sound of the waves.

THE EXTREME NORTHERN California coast is one of the most beautiful landscapes in the United States. The dramatic coastal scenes have been magnificently preserved by Redwood National Park and other adjoining state parks. Collectively, these world-class sanctuaries protect more than 30 miles of wild coastline south of the border with Oregon. It is in this region, just north of Crescent City, California, that the St. George Reef Lighthouse can be found.

The lighthouse, which was built on an exposed rock island six miles out to sea, was the most expensive lighthouse ever built in the country, costing over $700,000 and taking nearly ten years to build. One worker was swept out to sea during construction. In the end, the sturdy stone tower rose over 140 feet above sea level and held a first-order Fresnel lens that displayed a bright, flashing light.

The seas are often wild around the famous St. George Reef Lighthouse. At times the waves crash to the very base of the lighthouse structure, reminding us all of the frailty of human endeavor when placed beside the power of Mother Nature.

While its rock base is plainly visible during low tide, the lighthouse is often nearly submerged around its base by the violent Pacific storms that beset the coast every winter. The lighthouse was one of the most vital ones in northern California and southern Oregon, lighting a dark and dangerous area between Cape Blanco and Cape Mendocino. It was there, for example, that the *Brother Jonathan* struck ground off Point St. George in 1865 and went down with 215 passengers and crew. The lighthouse proved a dangerous place to work; several lighthouse keepers died when their boats swamped in the heavy surf. It was also one of the loneliest (no families permitted) and most unpleasant tours of duty in the service.

Finally, the remote lighthouse was decommissioned in 1975 and replaced by a floating light buoy. The historic St. George Reef Light undoubtedly saved many a mariner's life during its 83 years of operation. It represents—both in the difficulty of its construction and in the diligent attention to duty of its keepers—the very best in this nation's long commitment to maritime safety.

Heceta Head

Oregon (1894)

Visitors to the Heceta Head Light in Oregon are treated to this wonderful view of the rugged coast line, which evokes the mighty cliffs and redwood forests of the Big Sur Country farther to the south in California.

FROM THE BEGINNING, the west coast of the United States posed unique challenges to lighthouse builders. The coast is long and varied, as it ranges from sheer rock cliffs to heavily forested ridges to rolling dune fields. Violent winter storms and near-daily fogs arise in many areas. After the Civil War, as the nation's population and business interests surged westward, there was ever-increasing shipping to protect. Nowhere along that thousand-mile coast from Mexico to Canada did the Lighthouse Service find a greater challenge than along the wild Oregon coast, in particular the nearly 100-mile stretch between Cape Foulweather and Cape Argo.

It was there, in 1894, almost due west from Eugene, that the beautiful Heceta Head Light was built. The surrounding, near-vertical rock cliffs—which evoke the plunging coastlines of Big Sur, California—made construction of the light particularly difficult in the area. The white conical tower was raised on a flattened deck notched in the side of a mountain. Although the tower was only 56 feet tall, its actual height above the waves was 205 feet. The Heceta Head Light was, for its century, one of the more expensive lighthouses built, costing nearly $180,000.

The keeper's residence at the Heceta Head Light is as fine as anything in the system. The nearby forests are full of black-tailed deer and bright wildflowers.

Heceta Head, Oregon (1894)

In such a stunning setting, the picturesque light tower was destined to become a favorite of lighthouse aficionados and tourists, not to mention maritime painters and landscape photographers. Even by the off-the-scale beauty standards of northern Oregon, this is a lighthouse that has to be seen to be believed, with the cool green pine forests, black rock cliffs, and chromium blue ocean below. Few drivers can pass by on U.S. Highway 101 and not pull off to admire this awesome beacon. The lighthouse remains active today, and its light is said to be visible for over 20 miles out to sea.

The Heceta Head Light was automated in 1963 and is the brightest on the Oregon coast. The original lens was a first-order Fresnel.

The Heceta Head Light was named after Bruno de Heceta, who was an 18th-century Spanish Basque explorer of the Pacific coast. He was originally sent to explore the area by the Viceroy of New Spain to confirm whether or not colonial Russian and British settlements were appearing in the territory of Alta California. Heceta and his crew were the first Europeans to discover the mouth of the Columbia River flowing into the Pacific Ocean.

The original site of the Heceta Head Light contained numerous buildings, including two assistant keeper's quarters, the head lighthouse keeper's quarters, a farmhouse, and two kerosene storage buildings. After the light was electrified, the head lighthouse keeper's quarters was no longer needed so it was purchased for $10 and demolished for its lumber. The lumber was used to build a bookstore in nearby Mapleton.

The Heceta Head Light and its property is open to the public, and the still standing keeper's quarters are now used as a bed and breakfast.

Five Finger

Alaska (1902)

The nearby waters around the Five Finger Light abound with some of the greatest concentrations of animals in the world, including such rare species as the bald eagle and killer whale.

THE FABLED INSIDE PASSAGE of the Alaskan panhandle has been a favorite of maritime travelers since the days of John Muir. The distinguished naturalist wrote regularly of the scenic wonders of the coastal region for *Century* magazine in the late 19th century. Even at that time, the area was known for its dangerous shoals and reefs. (The northern waters became especially well known in this regard in 1989 after the grounding of the oil tanker *Exxon Valdez* on Bligh Reef.) For that reason, a lighthouse was placed on an island just north of Kupreanof Island in 1902. The area was known as "the five fingers" by shippers for the finger-like shape of the forested islands.

Located about 70 miles south of Alaska's state capital, Juneau, the Five Finger Light is often seen by those who take the Alaska Maritime Highway on scenic summer cruise vessels. The current lighthouse, with its art deco style, was completed in 1935, after the original wooden keeper's home and lighthouse burned down. Built of reinforced concrete, the new main building also serves as a sturdy platform for the nearly 70-foot tower. Interestingly enough, the Five Finger Light was one of the last to become automated (1984) in the United States.

Hopefully, the Five Finger Light will have a better fate than that of another famous Alaskan lighthouse. The Scotch Cap Lighthouse in the Aleutian Islands was swept out to sea by a tidal wave on April 1, 1946, along with its complement of five men. The Five Finger Light certainly serves a vital purpose, given the sheer volume of traffic along the Inside Passage, which includes not only ferries and cruise ships but also a steady stream of commercial crabbers and shrimpers—as well as numerous sport fishing vessels. The region truly is, as John Muir wrote, "a wonderland" of nature.

The lighthouse at Five Finger in Alaska is not as remote as one might think. Each summer, hundreds of boats ply the nearby waters, and visitors on the Inland Ferry and on cruise ships sail by at regular intervals.

Admiralty Head

Washington (1903)

WHIDBEY ISLAND IS fast becoming one of the most popular suburban areas in the greater Seattle area. Accessible by a short ferry ride from the burgeoning metropolis, the island offers city-dwellers a quiet, semirural refuge at the end of the work day. Here, those who work in the busy city across the waters can find relaxing pastoral scenes and an abundance of wildlife. It is not unusual to spot an orca, or killer whale, out on the saltwater, or see a bald eagle eating a salmon in a seaside tree. There are also great opportunities for those who love kayaking, boating, and fishing.

The island is also the site of four wonderful state parks, including Fort Casey State Park, which, among other things, protects the historic Admiralty Head Light. It was this light that helped to show mariners the way on their final transit of Puget Sound toward the piers, loading docks, and shipyards of Seattle and Tacoma.

It's a rare sunny day at the Admiralty Head Light on Whidbey Island near Seattle. The striking architectural design makes this lighthouse one of the most beautiful in the Pacific Northwest.

The initial light at Admiralty Head was commissioned by the United States Lighthouse Board in 1861. By that time, commercial traffic in Puget Sound had greatly increased as trade with the Far East opened up. The lighthouse consisted of a 41-foot stone tower complete with a fourth-order Fresnel lens. The station's light, which could be seen from about 15 miles away, was a welcome beacon to those coming in from the often rough Strait of Juan de Fuca. During the Spanish-American War, the Army dismantled the original lighthouse to make room for Fort Casey, an installation designed to help protect Seattle in the event of attack.

A new lighthouse was built in 1903. It included a short, thick stone tower attached to a two-story Spanish stucco residence that was painted a bright zinc white. Because of the height of Admiralty Head on Whidbey Island, the light was actually about 120 feet above the water. As Puget Sound shipping routes changed during the 1920s, there was less of a need for a lighthouse at Admiralty Head. Eventually, in 1922, the lighthouse was decommissioned, becoming, of all things, an officer's quarters for Fort Casey. Today it is a museum and popular tourist attraction. The lighthouse was recently featured by the U.S. Postal Service in its set of ten lighthouse stamps.

After the Admiralty Head Light was decommissioned in 1922, its fourth-order Fresnel Lens was moved to the then recently commissioned New Dungeness Lighthouse located on the Dungeness Spit in the Strait of Juan de Fuca.

Makapuu Point

Hawaii (1909)

Panoramic views of the Pacific greet visitors to the Makapuu Point Light. During the balmy winter months, groups of migratory whales are often spotted in nearby waters.

IN 1905, about seven years after the United States acquired the Hawaiian Islands, the U.S. Lighthouse Board stated the following: "All deep-sea commerce between Honolulu and Puget Sound, the Pacific coast of the United States, Mexico and Central America, including Panama, passes Makapuu Head, and...[yet] there is not a single light on the whole northern coast of the Hawaiian Islands to guide ships or warn them of their approach to land, after a voyage of several thousand miles."

One year later the U.S. Congress appropriated $60,000 to build a lighthouse at Makapuu Head, the easternmost point of Oahu Island. Because the site was nearly 400 feet above sea level, there was no need to build an extremely tall lighthouse. The lens, however, was enormous—eight and a half feet in diameter—so as to cast a strong beam far out to sea. The hyper-radiant lens, in fact, was the most powerful ever built and used in America.

Makapuu Point, located about 12 miles east of Honolulu at the terminus of the Koolau Range, rises 647 feet above sea level. At the tip of the volcanic point is the Makapuu Point Lighthouse, which offers a scenic view down onto the cerulean blue tropical waters and nearby white sand and black lava beaches. The highland area around the lighthouse is also popular with hang gliders, who take off directly from the rocky cliffs and soar like liberated birds above the lighthouse and sea.

Because of its unusual height, Makapuu Point Lighthouse offers one of the most breathtaking coastal panoramas of any in the United States. What's more, visitors to the famous lighthouse can see related maritime sites nearby, which include the Pacific Whaling Museum and Sea Life Park, Hawaii's only marine park.

The Hyper-Radiant Fresnel lens is nearly 12-feet tall. Many claim that the procedure used to make the lens is no longer known, making it a very unique lens for a very unique lighthouse.

Makapuu Point, Hawaii (1909)

Makapuu Point is a tremendously popular site on the Hawaiian island of O'ahu, where people can hike and take in fantastic coastal views.

Should the 1000 watt, 120 volt lamp burn out, there is a tangent lamp that will automatically rotate into place for operation.

Hawaii has few sites as popular as Makapuu Point, with its commanding prospect over the world. It is an ideal place to stop and recall the words of poet William Butler Yeats, who wrote of "time past and passing, time to come."